Branching in the Presence of Symmetry

D. H. SATTINGER
School of Mathematics
University of Minnesota

SOCIETY for INDUSTRIAL and APPLIED MATHEMATICS • 1983

PHILADELPHIA, PENNSYLVANIA 19103

Library of Congress Catalog Card Number 82-61451
ISBN 0–89871-182-7

Printed in Northern Ireland for the Society for Industrial and Applied Mathematics by The Universities Press (Belfast) Ltd.

Contents

Preface

I would like to express my appreciation to Joe McKenna for his efforts in organizing this CBMS conference at the University of Florida, Gainesville, Florida, which was held December 14–18, 1981. In addition to the support of the National Science Foundation, the meeting received financial support from the Mathematics Department and the Graduate Center of the University of Florida.

The subject of bifurcation theory has expanded so extensively in the last decade that only a limited survey of recent developments is feasible in these lectures. I have focussed on four areas: minimax methods for nonconvex functionals, equivariant bifurcation theory, equivariant singularity theory and critical orbits of group actions.

Two minimax theorems are presented in §1.1, the "mountain pass lemma" of Ambrosetti and Rabinowitz and the "linking spheres lemma," which, in the form presented here, is due to W. M. Ni. The linking spheres lemma says that if f is a smooth function on $\mathbb{R}^N$, if M^k and M^{N-k-1} are smooth manifolds in $\mathbb{R}^N$ with a nontrivial linking number, and if $f|_{M^k} \leqq \alpha < \beta \leqq f|_{M^{N-k-1}}$, then the number

$$c = \sup_{B^{N-k}} \operatorname{Min}_{x \in B^{N-k}} f(x),$$

where $\partial B^{N-k} = M^{N-k-1}$, is a critical value of f. There is a dual minimax problem, namely

$$c' = \inf_{S^k} \operatorname{Max}_{x \in S^k} f(x),$$

where S^k links M^{N-k-1} and $\partial S^k = 0$. These two critical values are the same. This is established in Theorem 1.2 by Alexander duality, based on a suggestion of R. Gulliver.

In §1.2 the linking spheres lemma is used to prove that the nonlinear wave equation

$$\begin{aligned} &\Box u + f(u) = u_{tt} - u_{xx} + f(u) = 0, \\ &u(0, t) = u(\pi, t) = 0, \\ &u(x, t+2\pi) = u(x, t) \end{aligned}$$

possesses nontrivial periodic solutions. This result was initially established by Rabinowitz by showing that a modified form of the nonconvex functional

$$J(u) = \int_0^{2\pi} \int_0^{\pi} \frac{1}{2}(u_x^2 - u_t^2) + F(u)\, dx\, dt$$

has nontrivial critical points. Rabinowitz's methods were refined and simplified considerably by Brezis, Coron and Nirenberg, who introduced a dual variational problem and applied the mountain pass lemma. A third method, based on the fact that the Fourier projection operators are bounded in L_p for $1<p<\infty$, is presented here. This approach mounts a direct attack on the functional J.

In §1.3 I have presented Clarke and Ekeland's elegant proof of the existence of 2π-periodic solutions of a Hamiltonian system with convex Hamiltonian H. Section 1.4 describes the application of isoperimetric problems in the treatment of free boundary value problems, and §1.5 gives a brief discussion of the topological methods of Bahri and Berestycki in treating perturbations of functionals which are invariant under a group action. The references mentioned in Chapter 1 are intended to give the highlights of the subject of critical point theory, but definitely do not constitute an exhaustive bibliography of the subject.

The basic notions of equivariant bifurcation theory have been presented in Chapter 2, but I have refined the presentation by introducing the notion of modules of equivariant maps over rings of invariant functions. This approach works in cases where one can compute the Hilbert basis for the ring of invariant functions and the generators of the module (not always possible). When it does work it is computationally more powerful than the traditional method of constructing the general covariant mappings of all orders. The module approach is at the foundation of the application of equivariant singularity theory which has been applied to bifurcation theory by Golubitsky and Schaeffer.

The presence of continuous symmetry groups (Lie groups) renders the implicit function theorem ineffective (as explained in §2.2), and a more refined technique is needed to obtain the full bifurcation picture. Equivariant singularity theory seems to be the right tool, though the computations involved can easily get out of hand.

In §2.3 I give a careful introduction to the representation $A \to OAO^+$ of $SO(3)$, where A is a symmetric traceless matrix and $O \in SO(3)$. This representation contains the representation $D^{(2)}$ and is fundamental to Golubitsky and Schaeffer's unfolding of the singularity for that representation. By restricting this representation to the subspace of diagonal traceless matrices, the branching problem is reduced to one with a discrete symmetry; that is, the problem is reduced to unfolding a singularity in the presence of a two-dimensional representation of S_3.

There is a close relationship between the presentation above and the adjoint representation of a Lie group on its algebra. In particular, the adjoint representation of $SU(3)$ which plays a fundamental role in the theory of strong interactions is $A \to UAU^*$. When this action is restricted to the Cartan subalgebra, the symmetry group is the same two-dimensional representation of S_3 as before; S_3 is the Weyl group for $SU(3)$. A brief account of the $SU(3)$ model and its role in particle physics is given in §2.5.

Sections 2.6 and 2.7 contain an account of the Hopf bifurcation theorem in the presence of spatial symmetries. The sections contain a streamlined version of the algebraic analysis, including a discussion of the stability calculations.

Chapter 3 contains a largely expository account of Golubitsky and Schaeffer's development of equivariant singularity theory. The basic notions of universal unfolding are discussed, and the unfolding of the S_3 singularity (which arises in the analysis of branching in the presence of rotational symmetry) is discussed in some detail.

Chapter 4 contains an account of L. Michel's theory of critical orbits of group actions. Michel has proved that certain vectors in the representation space, which are characterized as lying on orbits with a larger isotropy subgroup than any of their neighbors, are automatically *always* solutions of any equivariant bifurcation equation. A trace formula is given in §4.3 (using characters of the representation) for determining when such critical orbits exist. Specifically, if Γ is the representation and Δ is the isotropy subgroup, it is necessary and sufficient that Γ restricted to Δ contain the identity representation precisely once. In §4.4 this result is used to find critical orbits for the irreducible representations D^l of $O(3)$.

An interesting interrelationship between Michel's approach of critical orbits and equivariant singularity theory appears when one studies the unfolding of the $D^{(2)}$ singularity. When the codimension of the problem is zero, one obtains bifurcation only along the critical directions of symmetry breaking predicted by Michel's theory. When the codimension is not zero, however, the unfolding of the singularity leads to secondary bifurcation of solutions off the critical directions. This phenomenon is illustrated in detail by the examples.

I have omitted any detailed discussion of applications in this monogram, since I dealt with the applications of symmetry breaking in bifurcation in a previous survey article *Symmetry breaking and bifurcation in applied mathematics*, Bull. Amer. Math. Soc., 3 (1980), pp. 779–819. There the reader will find an extensive discussion of areas of application, as well as a bibliography of currently active fields.

These notes contain, in part, original research conducted with the support of the National Science Foundation under grant MCS 78-00415.

D. H. SATTINGER
University of Minnesota, Minneapolis
April, 1982

CHAPTER 1

Critical Points of Nonconvex Functionals

1.1. Minimax theorems. In this chapter we discuss techniques for finding critical points—in fact, saddle points—for functionals which are neither convex nor bounded above or below. Since we are not searching for global maxima or minima of these functionals, we cannot apply the classical direct methods of the calculus of variations. Instead, we resort to more sophisticated techniques.

The first technique, called the "mountain pass lemma", is quite intuitive. I will introduce it with a simple example (see [55]). Consider the functional

$$J(u)=\iint_D \frac{1}{2}(\nabla u)^2+\frac{u^3}{3}\,d\mathbf{x}$$

defined on the Hilbert space $\mathring{H}_1(D)$ (this is the closure of $C_0^\infty(D)$ under the Dirichlet norm). Here D is a bounded domain in $\mathbb{R}^n$. It is not hard to see that J has a local minimum at the origin. The second variation of J is

$$\delta^2 J(0)[\eta]=\frac{1}{2}\iint_D (\nabla\eta)^2\,d\mathbf{x}=\frac{1}{2}\|\eta\|^2,$$

and this term dominates the cubic term $\iint(\eta^3/3)\,d\mathbf{x}$ provided $n<6$. (This follows from the Sobolev inequality $|u|_p\leqq c\,\|u\|$ for $p<2n/(n-2)$.) On a small sphere about the origin, J is strictly positive. However, if u is a function such that $\iint u^3\,dx<0$ then, for all sufficiently large λ,

$$J(\lambda u)=\frac{\lambda^2}{2}\|u\|^2+\frac{\lambda^3}{3}\iint u^3\,dx<0.$$

Consider the task of getting out of the valley at $u=0$ by climbing as little as possible. We would pick a direction u along which J eventually becomes negative, and walk off along the ray λu. The highest level climbed in that direction is

$$\sup_\lambda J(\lambda u);$$

and, intuitively, the height of the pass leading out of the valley is

$$d=\inf_u\sup_\lambda J(\lambda u).$$

One expects (and is correct) that d is a critical value of J and that there is a vector u_0 such that $J(u_0)=d$ and $J'(u_0)=0$, where J' is the gradient of J. In the present case, the gradient of J is $J'(u)=\Delta u-u^2$. Thus one obtains a nontrivial critical point of J (assuming one can show that $d>0$). More generally, Ambrosetti and Rabinowitz [1] have proved the following:

THEOREM 1.1. *Let f be a C^1 functional on a Banach space B which satisfies the*

Palais–Smale condition. Assume there are points x_0 and x_1, and an open neighborhood Ω of x_0 which does not contain x_1, such that

$$f(x_0), f(x_1) < c_0 = \inf_{\partial\Omega} f.$$

Then the number

$$c = \inf_P \operatorname{Max}_{x \in P} f(x) \geqq c_0$$

is a critical value of f, the variations being taken over all continuous paths P joining x_0 to x_1.

f is said to satisfy the Palais–Smale condition if every sequence $\{x_n\}$ such that $f(x_n)$ is bounded and $f'(x_n) \to 0$ contains a convergent subsequence. By $f'(x_n)$ is meant the gradient of f, that is, $f'(x) \in B^*$ and

$$\frac{d}{dt} f(x + tv) = \langle v, f'(x) \rangle,$$

where $\langle \cdot, \cdot \rangle$ is the bilinear pairing between B and B^*.

The mountain pass lemma is geometrically intuitive. The following result is more sophisticated, and is an outgrowth of another variational argument of Rabinowitz.

THEOREM 1.2. *Let f be a C^1 function on $\mathbb{R}^N$ which satisfies the Palais–Smale condition, and let $\tilde{S}^k$ and $\tilde{S}^{N-k-1}$ be linked k- and $(N-k-1)$-closed chains in $\mathbb{R}^N$ $(\partial \tilde{S}^k = \partial \tilde{S}^{N-k-1} = 0)$ such that*

$$\alpha = \operatorname{Max}_{\tilde{S}^k} f < \operatorname{Min}_{\tilde{S}^{N-k-1}} f = \beta.$$

Then the following are critical values of f:

$$c_1 = \inf_{S^k} \operatorname{Max}_{x \in S^k} f \qquad \text{where } S^k \text{ links } \tilde{S}^{N-k-1} \text{ and } \partial S^k = 0,$$

$$c_2 = \sup_{B^{N-k}} \operatorname{Min}_{x \in B^{N-k}} f(x) \qquad \text{where } \partial B^{N-k} = \tilde{S}^{N-k-1},$$

$$c_3 = \sup_{S^{N-k-1}} \operatorname{Min}_{x \in S^{N-k-1}} f(x) \qquad \text{where } S^{N-k-1} \text{ links } \tilde{S}^k \text{ and } \partial S^{N-k-1} = 0,$$

$$c_4 = \inf_{B^{k+1}} \operatorname{Max}_{x \in B^{k+1}} f(x) \qquad \text{where } \partial B^{k+1} = \tilde{S}^k.$$

Furthermore, $c_2 = c_1 \leqq \alpha < \beta \leqq c_3 = c_4$.

Ni [47] and Nirenberg [48] proved that c_2 and c_4 are critical values of f; their results extended previous minimax results of Rabinowitz [53].

I will give a proof of Theorem 1.2 below. The following result gives a clearer picture of the geometric nature of the resulting critical points.

THEOREM 1.3. *Suppose $c_1 = c_2$ and there exist smooth surfaces S^k linking $\tilde{S}^{N-k-1}$ and B^{N-k} spanning $\tilde{S}^k$ such that*

$$f|_{S^k} \leqq c_1 = c_2 \leqq f|_{B^{N-k}}.$$

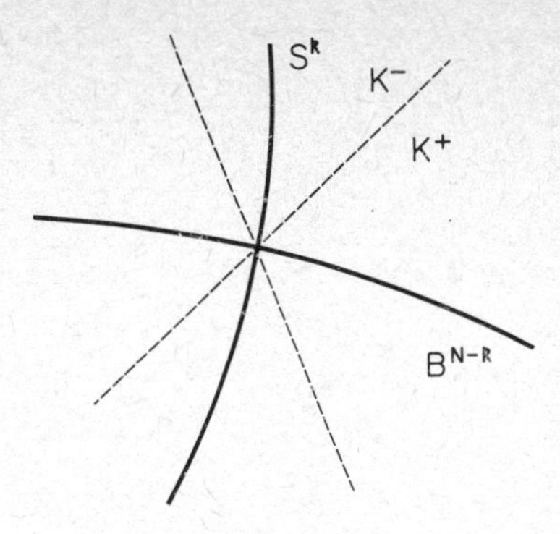

FIG. 1.1

Then $S^k \cap B^{N-k}$ *consists of critical points p at which* $f(p) = c_1 = c_2$. *If f is* C^2 *and p is a nondegenerate critical point (the Hessian* H_f *is nonsingular at p), then* H_f *has precisely k negative eigenvalues and* $N-k$ *positive eigenvalues, and* S^k *intersects* B^{N-k} *transversally.*

Let H be the Hessian of f and let

$$K_- = \{v \mid \langle Hv, v\rangle < 0\}, \qquad K_+ = \{v \mid \langle Hv, v\rangle > 0\}.$$

Then $T_p(S^k)$ *lies in* K_-, $T_p(B^{N-k})$ *lies in* K_+ *and* $k = \dim K_-$ *(see Fig. 1.1).*

The proof of Theorem 1.3 is relatively easy; those of Theorems 1.1 and 1.2 rest on the following deformation lemma, and Theorem 1.2 also requires an application of Alexander duality.

LEMMA 1.4. *Let f be a* C^1 *function on a Banach space B to* $\mathbb{R}$ *which satisfies the Palais–Smale condition and let*

$$A_s = \{x \mid f(x) \leqq s\},$$
$$B_s = \{x \mid f(x) \geqq s\},$$
$$K_c = \{c \mid f(x) = c,\ f'(x) = 0\}.$$

Let $\bar{\varepsilon} > 0$ *be given and suppose c is not a critical value of f (i.e.,* $K_c = \varnothing$*). Then for any* ε *in* $(0, \bar{\varepsilon})$ *there are families of Lipschitz homeomorphisms* φ_t *and* ψ_t, $0 \leqq t \leqq 1$, *such that*

1) $\varphi_t = \psi_t =$ *identity on* $A_{c-\varepsilon} \cup B_{c+\varepsilon}$,
2) $\varphi_1(A_{c+\varepsilon}) \subset A_{c-\varepsilon}$,
3) $\psi_1(B_{c-\varepsilon}) \subset B_{c+\varepsilon}$.

We shall give an example below that shows that the Palais–Smale condition is necessary even in one dimension.

Proof (see Palais, [50], p. 209]). If we have slightly more regularity on f, namely, if $f'(x)$ is Lipschitz, we can proceed as follows [48].

Construct a smooth $\eta(x)$ such that

$$\eta = 0 \quad \text{on } B_{c-\varepsilon} \cap A_{c+\varepsilon}, \qquad \eta = 0 \quad \text{on } A_{c-\bar{\varepsilon}} \cup B_{c+\bar{\varepsilon}},$$

and consider the flow

$$\dot{x} = -\eta(x)\frac{f'(x)}{\|f'(x)\|^2}, \qquad x(0) = y. \tag{1.1}$$

By the Palais–Smale condition, we may assume that $\|f'\|$ is bounded away from zero in $B_{c-\varepsilon} \cap A_{c+\varepsilon}$, as the flow is regular everywhere. Along the trajectories, $(d/dt)f(x(t)) = -\eta(x), \eta(x)$, so in time ε everything in $B_{c-\varepsilon} \cap A_{c+\varepsilon}$ flows into $A_{c-\varepsilon}$. Let $\hat{\varphi}$ denote this flow; that is, $\hat{\varphi}_t(y)$ is the solution of (1.1) at time t with initial data y. Then $\varphi_t = \hat{\varphi}_t$ is the required homotopy. ψ is obtained similarly, simply by changing the sign in (1.1).

The Palais–Smale condition is necessary for the deformation lemma even in one dimension. Consider the function f shown in Fig. 1.2. Any homeomorphism φ of $\mathbb{R}^1$ must be monotonic yet carry $A_{c+\varepsilon}$ into $A_{c-\varepsilon}$ and leave $B_{c+\varepsilon}$ invariant. Now $x_1 < x_2 < x_3$ and $\varphi(x_2) = x_2$; yet we must have $x_2 < \varphi(x_1)$ and $x_2 < \varphi(x_3)$, so φ cannot be monotonic. The Palais–Smale condition fails for f since we may choose $x_n \to -\infty$, where $f(x_n) \to c$ and $f'(x_n) \to 0$. To prove that $c_1 = c_2$ and $c_3 = c_4$ in Theorem 1.2 above, we need to apply Alexander duality. (This proof is due to R. Gulliver.)

ALEXANDER DUALITY (see Spanier [60, pp. 289, 295–6]). *Let A be compact in $\mathbb{R}^N$. Then $\tilde{H}_k(\mathbb{R}^N \backslash A; \mathbb{Z}) = \check{H}^{N-k-1}(A; \mathbb{Z})$, where $\tilde{H}_k$ is the relative homology of $\mathbb{R}^N \backslash A$ and $\check{H}^{N-k-1}(A; \mathbb{Z})$ is the direct limit of the cohomology classes $H^{N-k-1}(U; \mathbb{Z})$ where U are open neighborhoods containing A. The isomorphism γ between $H_k(\mathbb{R}^N \backslash A; Z)$ and $H^{N-k-1}(A; \mathbb{Z})$ may be defined in terms of linking numbers in the following way. Let K be a closed k-chain in $\mathbb{R}^N \backslash A$ ($\partial K = 0$) and let $[K]$ be the* k*th homology class containing K. Then $\gamma([K])$ is the element of $\check{H}^{N-k-1}(A; \mathbb{Z})$ defined by*

$$\gamma([K])([C]) = \textit{linking member of } C \textit{ with } K,$$

where C is a closed $(N-k-1)$-chain in A.

We use Alexander duality in this form to prove the following lemma.

LEMMA 1.5. *Let S^k be a closed k-chain in $\mathbb{R}^N$ and a a compact set in $\mathbb{R}^N$ with nonempty interior such that $A \cap S^k = \varnothing$ and A intersects every $(k+1)$-chain*

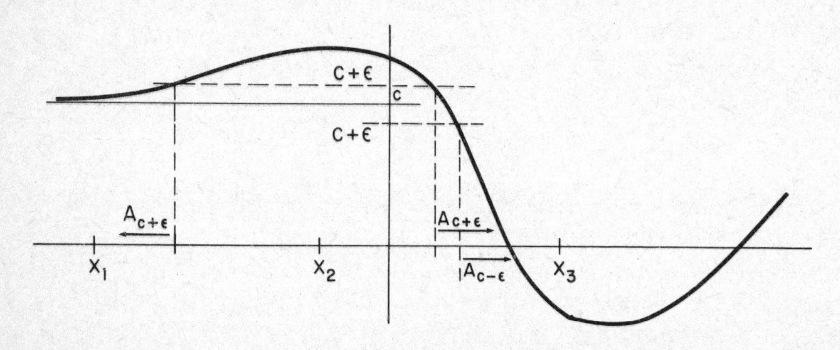

FIG. 1.2

B^{k+1} *such that* $\partial B^{k+1}=S^k$. *Then A contains a closed* $(N-k-1)$*-chain* S^{n-k-1}*that links* S^k.

Proof. If A does not contain such a chain, then $\gamma([S^k])([C])=0$ for all closed $(N-k-1)$-chains C in A. Therefore $\gamma([S^k])=0$ in $\check{H}^{N-k-1}(A;\mathbb{Z})$. Since γ is an isomorphism, $[S^k]=0$ in $\tilde{H}_k(\mathbb{R}^N\backslash A;\mathbb{Z})$. This means that S^k is the boundary of a chain B^{k+1} in $\mathbb{R}^N\backslash A$, but this contradicts the hypothesis of the lemma.

Proof of Theorem 1.2. Nirenberg and Ni proved that c_2 and c_4 are critical values; the proof that c_1 and c_3 are critical values is similar and goes like this. We first show that $c_1\geqq c_2$. If S^k links $\tilde{S}^{N-k-1}$ and B^{N-k} spans $\tilde{S}^{N-k-1}$, then S^k must interest B^{n-k}, so

$$\operatorname*{Max}_{x\in S^k} f\geqq \operatorname*{Min}_{x\in B^{n-k}} f(x).$$

Minimizing over S^k and maximizing over B^{N-k} we get $c_1\geqq c_2$.

We now show c_1 is a critical value. Given any $\varepsilon>0$ there is, by definition of c_1, a linking S^k such that $\operatorname{Max}_{x\in S^k} f(x)<c_1+\varepsilon$. We may rephrase this by saying that for every $\varepsilon>0$ the set $A_{c_1+\varepsilon}$ continues a linking chain S^k. We choose ε so that $c_1+\varepsilon<\beta$. If c_1 is not a critical value, there exists a family of homeomorphisms φ_t such that $\varphi_1(A_{c_1+\varepsilon})\subset A_{c_1}-\varepsilon$ and φ_t acts as the identity on $B_{c_1+\varepsilon}$. Since $\operatorname{Min}_{x\in\tilde{S}^{N-k-1}} f(x)\geqq\beta>c_1+\varepsilon$, φ_t acts as the identity on $\tilde{S}^{N-k-1}$. Therefore S^k is homotopic to a chain $S'^k=\varphi_1(S^k)$ which also links $\tilde{S}^{N-k-1}$, and $S'^k\subset A_{c_1-\varepsilon}$. This means that $\operatorname{Max}_{x\in S'^k} f(x)<c_1-\varepsilon$, which contradicts the definition of c_1. Therefore c_1 is a critical value.

To show $c_1\leqq c_2$, it suffices to prove that $c_2>c_1-\varepsilon$ for any $\varepsilon>0$. To do this we must show that every set $\mathring{B}_{c_1-\varepsilon}=\{x\mid f(x)>c_1-\varepsilon\}$ contains an $(N-k)$-chain B^{n-k} which spans $\tilde{S}^{N-k-1}$. Now let us suppose that every spanning chain meets $(\mathring{B}_{c_1-\varepsilon})^c=A_{c_1-\varepsilon}$. Now $A_{c_1-\varepsilon}$ is a closed set which does not meet $\tilde{S}^{N-k-1}$. If $A_{c_1-\varepsilon}$ is compact, then by the lemma $A_{c_1-\varepsilon}$ contains a k-chain S^k which links $\tilde{S}^{N-k-1}$. But this contradicts the definition of c_1. To deal with the case where $A_{c_1-\varepsilon}$ is not compact, replace f by a function $\hat{f}$ such that $\hat{f}=f$ for $|x|\leqq R$ and $\hat{f}\to\infty$ as $|x|\to\infty$. If R is sufficiently large, the critical values c_1 and c_2 for $\hat{f}$ will be the same as for f. We assume R is chosen so large that there are no critical points of f outside B_R with critical values lying in some fixed interval about c_1. (This is possible by the Palais–Smale condition.) Then outside $B_R=\{x\mid |x|\leqq R\}$ we can apply the deformation lemma to carry the set $A_{c_1+\varepsilon}$ into $A_{c_1-\varepsilon}$. That means that any S_k which intersects B_R^c can be deformed in B_R^c so that $f|_{S^k\cap B_R^c}<c_1-\varepsilon$. So c_1 may be obtained by taking the infimum over all S^k which lie inside B_R; hence $c_1(f)=c_1(\hat{f})$. Similarly $c_2(f)=c_2(\hat{f})$. But since $\hat{f}\to\infty$ as $|x|\to\infty$, $\hat{A}_{c_1-\varepsilon}$ is compact for $\hat{f}_1$, so the above argument goes through and $\hat{c}_1=\hat{c}_2$.

1.2. Periodic solutions of a nonlinear wave equation. A minimax argument such as that of Theorem 1.2 in the preceding section was used by Rabinowitz [53] to prove the existence of nontrivial periodic solutions of the nonlinear wave equation

$$\Box u+f(u)=u_{tt}-u_{xx}+f(u)=0, \tag{1.2}$$

$$u(0,t)=u(\pi,t)=0, \qquad u(x,t+2\pi)=u(x,t). \tag{1.3}$$

Rabinowitz applied minimax arguments to demonstrate the existence of critical points of the functional

$$J(u)=\int_0^{2\pi}\int_0^{\pi}\frac{1}{2}(u_x^2-u_t^2)+F(u)\,dx\,dt.$$

His methods were quite complicated, but the problem was given a simple, more refined treatment by Brezis, Coron and Nirenberg [14]. We present a third method, based on a direct application of Theorem 1.2, and then indicate the approach used in [14].

In what follows we denote by $H(u, v)$ the quadratic form

$$H(u,v)=\int_0^{2\pi}\int_0^{\pi}(u_xv_x-u_tv_t)\,dx\,dt.$$

Concerning the nonlinear term f and $F(u)=\int_0^u f(s)\,ds$, we assume that f is a smooth, nondecreasing function, that $f(0)=f'(0)=0$, that $f(u)/u\to+\infty$ as $|u|\to\infty$, and that there exist constants $p>2$ and $\beta>2$ such that

$$0<\varliminf_{|u|\to\infty}\frac{|f(u)|}{|u|^{p-1}}\leqq\varlimsup_{|u|\to\infty}\frac{|f(u)|}{|u|^{p-1}}<+\infty \tag{1.4}$$

and

$$\frac{uf(u)}{F(u)}\geqq\beta. \tag{1.5}$$

Condition (1.5) has a simple geometric interpretation, as shown in Fig. 1.3. The quantity $uf(u)/2$ is the area under the triangle with vertex at $(u, f(u))$, while $F(u)$ is the area under the graph of $f(u)$.

As a consequence of (1.4), F behaves like $|u|^p$ at ∞, and there exist constants a and b such that

$$a\,|u|^p\leqq F(u)\leqq b\,|u|^p. \tag{1.6}$$

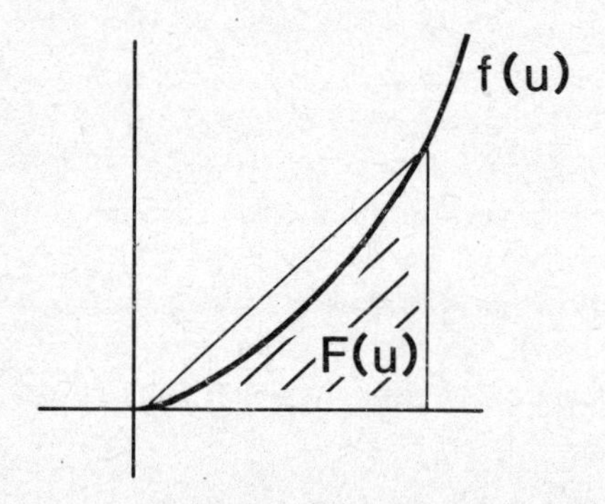

FIG. 1.3

Periodic functions on $\Omega=(0,2\pi)\times(0,\pi)$ may be expanded in a Fourier series

$$u=\sum_{j,k} A_{jk}e^{i(jx+kt)}$$

where the A_{jk} satisfy certain symmetry conditions in order that u be real and satisfy the boundary and periodicity conditions (1.3), namely $A_{jk}=-A_{j,-k}=-A_{-j,k}=A_{-j,-k}$. The kernel $N=N(\Box)$ is formally spanned by the Fourier modes for which $j^2=k^2$. We define also the subspaces

$$R^{\pm}=R^{\pm}(\Box)=\text{Span}\,\{e^{i(jx+kt)}\mid(j^2-k^2)>0\}.$$

We put the L^2 topology on these spaces. Our quadratic form $H(u,u)$ is zero on N, positive definite on R^+ and negative definite on R^-: for

$$H(u,u)=4\pi^2\sum_{j,k}(j^2-k^2)\,|A_{jk}|^2.$$

Let $R=R^+\oplus R^-$; clearly $R=N^\perp$. We have

LEMMA 1.6. *$\Box$ restricted to $R^+(R^-)$ is positive (negative) definite; and, for $1\leqq p<\infty$, there is a constant c such that*

$$|u|_p\leqq c\sqrt{-H(u,u)},\qquad u\in R^-. \tag{1.7}$$

For $0\leqq s\leqq 1$, $\Box^{-s}$ is a bounded operator from R into $W_{s,2}\cap R$.

One also has the a priori estimate

$$\|f\|_{0,\alpha}\leqq c\,|\Box f|_p,\qquad \alpha=1-\frac{1}{p}, \tag{1.8}$$

where $\|\cdot\|_{0,\alpha}$ is the Hölder norm with exponent α on Ω.

Proof. Inequality (1.8) is derived from an explicit integral representation for $\Box^{-1}$ (see [14]). The multiplier for $\Delta^{s/2}\Box^{-s}$ on R^+ or R^- is

$$\frac{(j^2+k^2)^{s/2}}{(j^2-k^2)^s}\leqq\frac{(|j|+|k|)^s}{(|j|+|k|)^s\,|\,|j|-|k|\,|^s}=\frac{1}{|\,|j|-|k|\,|^s}\leqq 1.$$

Therefore $\Delta^{s/2}\Box^{-s}$ is a bounded transformation on R; consequently on R, $\|u\|_{s,2}=|\Delta^{s/2}u|_2\leqq|\Box^s u|_2$ where $\|\cdot\|_{s,2}$ is the norm on $W_{s,2}$. This shows that $\Box^{-s}$ is a bounded mapping from R to $W_{s,2}\cap R$.

In two dimensions, for $1<p<\infty$ we have $|u|_p\leqq c\,\|u\|_{s,2}$; therefore on R^+ $|u|_p\leqq c\,|\Box^{1/2}u|_2=c\sqrt{H(u,u)}$. Similarly, (1.7) holds on R^-.

We use (1.7) to set up the minimax argument for the functional J. Let P_n be the projection of elements in $L^2(\Omega)$ onto Fourier modes with $|j|$, $|k|\leqq n$, let $Q_n=I-P_n$, and let E_n be the range of P_n. Restricting J to E_n, we get a smooth function on a finite dimensional vector space. Now

$$J(u)=\frac{1}{2}H(u,u)+\int_0^{2\pi}\int_0^{\pi}F(u)\,dx\,dt.$$

On $P_n(R^+\cup N)$, $J\geqq 0$, while on P_nR^-, J tends to $+\infty$ along any ray (see Fig.

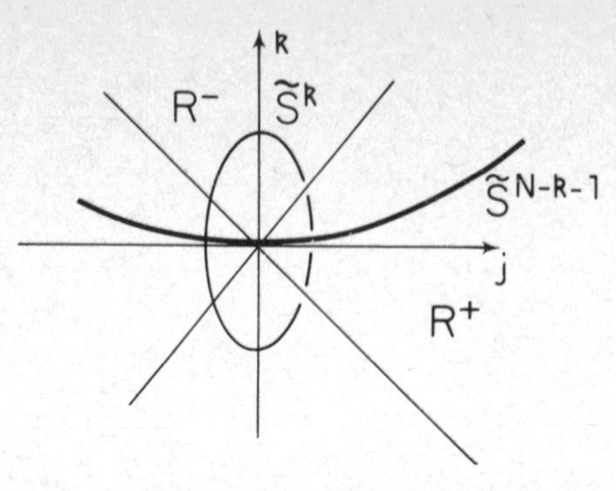

FIG. 1.4

1.4). In fact, in finite dimensions $|H(u, u)| \leqq (n^2-1)\,|u|_2^2$. On the other hand, the growth condition (1.6) implies that

$$\iint F(u)\,dx\,dt \geqq a \iint |u|^p\,dx\,dt = a\,|u|_p^p \geqq a'\,|u|_2^p.$$

Therefore

$$J(u) \geqq a'(|u|_2)^p - (n^2-1)(|u|_2)^2;$$

and, since $p>2$, $J(u)>0$ for sufficiently large $|u|_2$.

By (1.6) and (1.7),

$$\begin{aligned} J(u) &= \frac{1}{2} H(u, u) + \iint F(u)\,dx\,dt \\ &\leqq \frac{1}{2} H(u, u) + b\,|u|_p^p \\ &\leqq \frac{1}{2} H(u, u) + c(-H(u, u))^{p/2}. \end{aligned}$$

Putting $H(u, u) = -d^2$ we see that $J(u) < -d^2/2 + cd^p$. The right side assumes a minimum of $q = ((2-p)/2p)(1/cp)^{2/p-2}$, $q<0$, when $d = (1/cp)^{1/p-2}$. So in R^- we take the sphere given by $H(u, u) = -d^2$ to be the closed chain $\tilde{S}^k$ required in Theorem 1.2.

For the linking manifold S^{N-k-1} we take a sphere which passes through the origin lying in the cone R^+ and closes far enough out in R^- so that $J \geqq 0$. Then the conditions of Theorem 1.2 are satisfied and we have a critical value c_n given by

$$c_n = \inf_{S^k} \operatorname*{Max}_{u \in S^k} J(u) \leqq q < 0.$$

Note that the number q is fixed independently of the approximation E_n, so we have a uniform upper bound on the c_n's. A uniform lower bound on the c_n's may be obtained easily using a technique introduced by Rabinowitz in his original paper. For any ball B^{N-k} that spans $\tilde{S}^{N-k-1}$ we have

$$c_n \geqq \operatorname*{Min}_{u \in B^{N-k}} J(u).$$

Let us construct $\tilde{S}^{N-k-1}$ so that it lies in the subspace $R^+ \oplus N \oplus [\sin x \sin 2t]$. The minimum of J on B^{N-k} then occurs at a point $\bar{u} = au_{12} + w$, where $u_{12} = \sin x \sin 2t$ and $w \in R^+ \oplus N$. Therefore

$$c_n \geqq J(au_{12}+w) = \frac{1}{2} H(au_{12}+w, au_{12}+w) + \iint F(au_{12}+w)\, dx\, dt$$

$$\geqq -\frac{3a^2}{2} + \frac{1}{2} H(w, w) \geqq -\frac{3a^2}{2} |u_{12}|_2^2.$$

But $|u|^2 = a^2 |u_{12}|^2 + |w|^2 \geqq |u_{12}|_2^2$, so $c_n \geqq -\frac{3}{2} |\bar{u}|^2$.

Now we need an estimate on $|\bar{u}|^2$. On $R^+ \oplus N \oplus [u_{12}]$, $H(u, u) \geqq -3 |u|_2^2$; and at the mimimum point $\bar{u}$,

$$0 > q > c_n \geqq \frac{1}{2} H(\bar{u}, \bar{u}) + \iint F(\bar{u})\, dx\, dt,$$

so

$$-\iint F(\bar{u})\, dx\, dt \geqq \frac{1}{2} H(\bar{u}, \bar{u}) \geqq -\frac{3}{2} |\bar{u}|_2^2,$$

or

$$\frac{3}{2} |\bar{u}|_2^2 \geqq \iint F(\bar{u})\, dx\, dt.$$

But by (1.6)

$$\frac{3}{2} |\bar{u}|_2^2 \geqq \iint F(\bar{u})\, dx\, dt \geqq a\, |\bar{u}|_p^p \geqq C\, |\bar{u}|_2^p,$$

and, since $p > 2$, we get a uniform bound on $|\bar{u}|_2$. Our argument so far has followed that in Nirenberg's review article [48].

Our task now is to select a convergent subsequence from the initial points u_n. Let us write $u_n = w_n + \chi_n$ where $\chi_n \in N$ and $w_n \in R$. The critical points satisfy $\Box w_n = P_n f(u_n)$. Now

$$c_n = J(u_n) = \frac{1}{2} H(u_n, u_n) + \iint F(u_n)\, dx\, dt,$$

and

$$J'(u_n)(u_n) = H(u_n, u_n) + \iint f(u_n) u_n\, dx\, dt = 0.$$

Thus

$$c_n = J(u_n) - \frac{1}{2} J'(u_n)[u_n] = \iint F(u_n) - \frac{1}{2} u_n f(u_n)\, dx\, dt.$$

From (1.5), $uf(u)/2 - F(u) \geqq (\beta/2 - 1)F(u)$, where $\beta/2 - 1 > 0$; we therefore obtain a uniform bound on the integrals $\iint_\Omega F(u_n)$ and hence, effectively, on the L_p norms of the u_n.

Now let $G(u)$ be the conjugate convex function to F; F and G are related by Young's identity

$$F(u)+G(f(u))=uf(u).$$

Since the integrals $\iint_\Omega F(u_n)\,dx\,dt$ are bounded, so are $\iint_\Omega u_n f(u_n)$, and therefore also $\iint_\Omega G(f(u_n))$. Moreover, since $F\sim u^p$, $G(u)\sim u^{p'}$ where $1/p+1/p'=1$. Therefore the functions $f(u_n)$ are uniformly in $L_{p'}$. From the equations $\Box w_n = P_n f(u_n)$ we may then obtain uniform estimates on $\|w_n\|_{0,\alpha}$, where $\alpha=1-1/p'$, by virtue of (1.8). In fact, since the norms $|f(u_n)|_{p'}$ are uniformly bounded, so are $|P_n f(u_n)|_{p'}$. This follows from the fact that the projections P_n are uniformly bounded in $L_{p'}$. (See [64, p. 153]. Since, for $p>1$, $|P_n f-f|_p\to 0$, the family $\{P_n f\}$ is bounded in n for every $f\in L_p$. Therefore, by the uniform boundedness principle the projections P_n are uniformly bounded in the L_p norm.)

We now choose a subsequence of the u_n (which we continue to denote by u_n) such that $u_n \rightharpoonup u$ in L_p, w_n is strongly convergent in $C_{0,\alpha'}$ for $\alpha'<\alpha$, and $\Box w_n$ is weakly convergent in L_2. From the convexity of F,

$$\begin{aligned}\iint F(u)-F(u_n) &\geqq \iint f(u_n)(u-u_n)\\ &= \iint f(u_n)(w-w_n)+\iint f(u_n)(\chi-\chi_n).\end{aligned}$$

Now $w_n\to w$ in L_p and $|f(u_n)|_{p'}<C$, so the first term above on the right goes to zero. Moreover,

$$\iint f(u_n)\chi_n=\iint P_n f(u_n)\chi_n=\iint \Box w_n\chi_n=0,$$

and

$$\begin{aligned}\iint f(u_n)\chi &= \iint P_n f(u_n)\chi+\iint Q_n f(u_n)\chi\\ &= \iint (\Box w_n)\chi+\iint Q_n f(u_n)\chi\\ &= \iint Q_n f(u_n)\chi=\iint f(u_n)Q_n\chi.\end{aligned}$$

But $|f(u_n)|_{p'}$ is bounded and $|Q_n\chi|_p\to 0$. Therefore

$$\iint F(u)\geqq \overline{\lim_{n\to\infty}}\iint F(u_n).$$

Finally, we show that in fact $u_n\to u$ strongly in L_p. This step depends on the following lemma:

LEMMA 1.7. *If* (1.6) *is satisfied, there exists a positive constant c such that*

$$F(y)-F(x)\geqq f(x)(y-x)+cF(y-x), \tag{1.9}$$

$$G(y)-G(x)\geqq g(x)(y-x)+cG(y-x), \tag{1.10}$$

where G is the conjugate convex function.

Proof.

$$F(y)-F(x)-f(x)(y-x)=\int_x^y f(t)-f(x)\,dt.$$

To prove the result it suffices to show there exists a constant $c>0$ such that

$$\frac{f(t)-f(x)}{f(t-x)} \geqq c \quad \text{for all } x, t.$$

The above quotient is positive everywhere and, since $f(0)=f'(0)=0$, it tends to $+\infty$ as $t \to x$. On the other hand, from (1.4),

$$\begin{aligned}\varliminf_{|t|\to\infty} \frac{f(t)-f(x)}{f(t-x)} &= \varliminf_{|t|\to\infty} \frac{|f(t)|}{|t|^{p-1}} \cdot \frac{|t-x|^{p-1}}{|f(t-x)|} \\ &\geqq \varliminf_{|t|\to\infty} \frac{|f(t)|}{|t|^{p-1}} \cdot \varliminf_{|t|\to\infty} \frac{|t-x|^{p-1}}{|f(t-x)|} \\ &= \varliminf_{|t|\to\infty} \frac{|f(t)|}{|t|^{p-1}} \left(\varlimsup_{|t|\to\infty} \frac{|f(t-x)|}{|t-x|^{p-1}} \right)^{-1} > 0.\end{aligned}$$

Note that the last step requires that $\varlimsup_{|t|\to\infty} (|f(t)|/|t|^{p-1}) < +\infty$. Inequality (1.10) is proved in the same way. Inequalities (1.4) for f imply similar inequalities for $g=f^{-1}$.

Now applying (1.9) to the sequence u_n we get

$$\iint F(u_n)-F(u) \geqq \iint f(u)(u_n-u)+c\iint F(u_n-u).$$

Since $f(u)\in L_{p'}$ and $u_n - u \rightharpoonup 0$ in L_p, the first term tends to zero. Therefore

$$0 \geqq \varlimsup_{n\to\infty} \iint F(u_n)-F(u) \geqq c \varlimsup_{n\to\infty} \iint F(u_n-u),$$

and hence $u_n \to u$ strongly in L_p.

We now prove that u is a critical point of J. We must show that for any $\varphi \in W_{1/2,2}$,

$$H(u,\varphi)+\iint f(u)\varphi\,dx\,dt=0.$$

Now

$$H(u,\varphi)+\iint f(u)\varphi = H(u,P_n\varphi)+\iint f(u)P_n\varphi + H(u,Q_n\varphi)+\iint f(u)Q_n\varphi.$$

$\|Q_n\varphi\|_{1/2,2}=|\Delta^{1/2}Q_n\varphi|_2=|Q_n\Delta^{1/2}\varphi|_2 \to 0$ as $n\to\infty$ and since $f(u)\in L_{p'}$, the third and fourth terms above go to zero. On the other hand, for fixed n

$$H(u,P_n\varphi)+\iint f(u)P_n\varphi = \lim_{k\to\infty} H(w_k,P_n\varphi)+\iint f(u_k)P_n\varphi = 0.$$

Recall that $w_k \rightharpoonup w$ in $W_{1,2}$, so that $H(w_n, P_n\varphi)\to H(u,P_n\varphi)$. (We leave it to the reader to prove that $f(u_k)\to f(u)$ in $L_{p'}$ when $u_k \to u$ in L_p.)

We must still prove that u is a nontrivial solution. We do this by showing that $J(u) \leqq q < 0$, where $q \leqq \overline{\lim}_n J(u_n)$. First, we have

$$\iint F(u) \geqq \overline{\lim_{n \to \infty}} \iint F(u_n)$$

from the argument preceding Lemma 1.6. On the other hand, since F is convex, we have by lower semicontinuity

$$\underline{\lim_{n \to \infty}} \iint F(u_n) \geqq \iint F(u),$$

and therefore

$$\iint F(u) = \lim_{n \to \infty} \iint F(u_n).$$

Note that this result applies even if u_n converges only weakly to u.

Now consider the quadratic term $H(u_n, u_n)$. We may write this as

$$\begin{aligned} H(w_n + \chi_n, w_n + \chi_n) = H(w_n, w_n) &= H(w_n, w + (w_n - w)) \\ &= (\Box w_n, w) + (\Box w_n, w_n - w). \end{aligned}$$

Now we recall that $|\Box w_n|_{p'} = |P_n f(u_n)|_{p'} \leqq \text{const.}$ and $w_n \to w$ strongly in L_p. Therefore the second term tends to zero. For the first term we have $(\Box w_n, w) = (w_n, \Box w) \to (w, w)$. Therefore

$$H(u_n, u_n) \to H(w, w) = H(u, u),$$

and

$$J(u) = \lim_{n \to \infty} J(u_n) \leqq q < 0.$$

This argument applies even if χ_n converges to χ only weakly in L_p.

We have shown:

THEOREM 1.8. *Under the assumptions* (1.4) *and* (1.5) *on* f, *the functional* J *has a nontrivial critical point* u. *This vector* u *is a weak* 2π *time periodic solution of the nonlinear hyperbolic equation* (1.2).

The regularity and boundedness of this weak time periodic solution has been demonstrated by Rabinowitz and also by Brezis et al. (see also [15]). The approach I have given here mounts a direct attack on the functional J. In order to obtain the necessary compactness results, Rabinowitz worked with the modified functional

$$J_\varepsilon(u) = \iint \frac{u_x^2 - u_t^2}{2} + \varepsilon \chi_t^2 + F(u)\, dx\, dt$$

where χ is the projection of u on the null space N. He then showed that the family $\{u_\varepsilon\}$ of critical points is equicontinuous as $\varepsilon \to 0$. Brezis et al. replace the functional J by a dual functional Q as follows. Say $\Box u + f(u) = 0$, and write $u = w + \chi$ where $w \in R$ and $\chi \in N$. Putting $v = f(u)$ we have $\Box w + v = 0$, hence $w + Av = 0$ where $A = \Box^{-1}$. Letting $g = f^{-1}$ we have $u = g(v) = w + \chi$, and therefore the equation formally takes the form

$$Av+g(v)=\chi \tag{1.11}$$

where $\chi \in N$. The functional for this equation is

$$Q(v)=\frac{1}{2}(Av, v)+\iint G(v)\,dx\,dt,$$

where $G(v)=\int_0^v g(s)\,ds$ is the conjugate convex function to F. The vector $\chi \in N$ in (1.11) is then a Lagrange multiplier for the problem. Critical points of Q are sought in the space R with the norm $L_{p'}$.

Brezis et al. show that Q satisfies the Palais–Smale condition when $f(u)=c\,|u|^{p-1}$ for some $p>1$. They do not prove the Palais–Smale condition in the nonhomogeneous case, but instead prove a weaker condition. We note, however, that the PS condition can still be shown to hold in the nonhomogeneous case provided (1.4) and (1.5) are satisfied; one simply applies the corresponding argument based on Lemma 1.6. (In fact, the argument here is a simple sharpening of theirs.) The demonstration of critical points of Q is based on the mountain pass theorem rather than on Theorem 1.2.

I chose to deal directly with the functional J for two reasons: (i) it illustrates an application of the linking chain lemma (Theorem 1.2), and (ii) the approach may be valuable in other contexts where the duality method cannot be used, for example, if f is not monotone. The direct approach used here made use of the fact that Fourier projections P_n converge strongly to the identity in L_p for $1<p<\infty$.

Brezis, Coron and Nirenberg [14] derive the existence of critical points under weaker assumptions on the nonlinear term f than those given here. Namely, they assume that: (i) $u^{-1}f(u)\to\infty$ as $|u|\to\infty$, and (ii) there exist constant $\alpha>0$ and $c>0$ such that

$$\tfrac{1}{2}uf(u)-F(u)\geqq \alpha\,|f(u)|-C \quad \text{for all } u.$$

We showed that the sequence of trial solutions contained a strongly convergent subsequence whose limit was a nontrivial critical point of the functional J. This proof depended on the properties (1.4) and (1.5) of f; (1.4) implies, among other things, that f is strictly monotone. Strong convergence followed on the basis of Lemma 1.6 from the fact that

$$\overline{\lim_{n\to\infty}}\iint F(u_n)\leqq \iint F(u).$$

Since, by lower semicontinuity,

$$\iint F(u)\leqq \underline{\lim_{n\to\infty}}\iint F(u_n),$$

we have

$$\lim_{n\to\infty}\iint F(u_n)=\iint F(u).$$

If F is not strictly convex, it does not necessarily follow that u_n converges strongly to u. For example, if $F(u)=|u|$ and u_n is a sequence of nonnegative functions which converges weakly to u, then

$$\iint F(u_n)=\iint u_n \to \iint u=\iint F(u)$$

even though u_n may only converge weakly to u.

In such cases it may still be possible to prove that the weak limit of sequence is a weak solution of the associated nonlinear problem by a trick of Minty's. In the present case, Minty's device proceeds roughly as follows:

From the monotonicity of f,

$$\iint (f(u_n)-f(\psi))(u_n-\psi)\geqq 0$$

for all $\psi\in L_p$. Therefore

$$\iint (P_nf(u_n)+Q_nf(u_n)-f(\psi))(w_n+\chi_n-\psi)\geqq 0,$$

$$\iint (-\Box u_n+Q_nf(u_n)-f(\psi))(w_n+\chi_n-\psi)\geqq 0.$$

Now $|Q_nf(u_n)|_{p'}\to 0$, $\iint (u_n)\chi_n=0$, $w_n\to w$ strongly, and $u_n\to u$ weakly. Therefore in the limit we obtain

$$\iint (-\Box u-f(\psi))(u-\psi)\geqq 0.$$

Choosing $\psi=u+t\eta$ for $\eta\in L_p$ and $t\geqq 0$, we get the inequality

$$\iint (\Box u+f(u+t\eta))\eta\geqq 0$$

for all η in L_p. Letting $t\to 0+$ we obtain

$$\iint (\Box u+f(u))\eta\geqq 0.$$

Since this is true for all $\eta\in L_p$, we may conclude that u is a weak solution of $\Box u+f(u)=0$. The proof that u is a nontrivial solution proceeds as before.

I would like to thank H. Brezis and C. Kenig for their helpful comments concerning some of the material in this section.

1.3. Periodic solutions of Hamiltonian systems. Rabinowitz [54] has also used variational methods to construct time periodic solutions of the Hamiltonian equations

$$\dot{x}=\frac{\partial H}{\partial p},\qquad \dot{p}=-\frac{\partial H}{\partial x} \tag{1.12}$$

where $H=H(x, p)$ and $x, p \in \mathbb{R}^n$. He proved the existence of orbits of period T by considering, essentially, the functional

$$K(x, p)=\int_0^T x \cdot \dot{p}+H(x, p)\, dt$$

where $x \cdot \dot{p}$ denotes the dot product in $\mathbb{R}^n$. His proof is again quite technical, but Clarke and Ekeland have shown that when H is convex, K may be replaced by a dual functional which is much easier to treat. Let me indicate the essential details here (following Ekeland [27], see also [28]).

If H is convex, the conjugate convex function G may be defined by Fenchel's formula:

$$G(y, q)=\sup_{(x,p)}\{x \cdot y+p \cdot q-H(x, p)\}.$$

If H is strictly convex, then we actually have the equality

$$G(y, q)+H(x, p)=x \cdot y+p \cdot q. \tag{1.13}$$

Here, x, y, p and q belong to $\mathbb{R}^n$, and $x \cdot y$ is the dot product in $\mathbb{R}^n$. Differentiating with respect to x, y, p, q we get the transformation formulae

$$y=\frac{\partial H}{\partial x}, \qquad q=\frac{\partial H}{\partial p}, \tag{1.14}$$

$$x=\frac{\partial G}{\partial y}, \qquad p=\frac{\partial G}{\partial p}. \tag{1.15}$$

LEMMA 1.9. *Consider the functional*

$$J(\dot{y}, \dot{q})=\int_0^1\{G(-\dot{q}, \dot{y})+Ty\dot{q}\}\, dt$$

defined on the space of functions $\dot{y}$, $\dot{q}$ *where* $\dot{y}$ *and* $\dot{q} \in L^{\alpha'}(0, 1)$ $(1<\alpha'<2)$ *which satisfy the boundary conditions* $y(0)=y(1)=0$, $q(0)=q(1)=0$. *We take* $y(t)=\int_0^t \dot{y}(s)\, ds$, *etc. Let* $\{y, q\}$ *be an extremal of* J *and define* x *and* p *by*

$$x(t)=G_1'\left(-\dot{q}\left(\frac{t}{T}\right), \dot{y}\left(\frac{t}{T}\right)\right),$$

$$p(t)=G_2'\left(-\dot{q}\left(\frac{t}{T}\right), \dot{y}\left(\frac{t}{T}\right)\right).$$

Then $(x(t), p(t))$ *is a* T*-periodic solution of* (1.12).

Proof. J is extremized over functions $\dot{y}$, $\dot{q}$ with mean zero, so the extrema satisfy the Euler–Lagrange equations

$$G_2'(-\dot{q}, \dot{y})-Tq+\mu=0, \qquad -G_1'(-\dot{q}, \dot{y})+Ty+\mu'=0$$

where μ and μ' are constants. Now put

$$x(t)=\mu'+Ty\left(\frac{t}{T}\right) \quad \text{and} \quad p(t)=Tq\left(\frac{t}{T}\right)\mu.$$

Then $\dot{x}(t)=\dot{y}(t/T)$, $\dot{p}(t)=\dot{q}(t/T)$, and

$$x(t)=G_1'(-\dot{p}(t), \dot{x}(t)),$$
$$p(t)=G_2'(-\dot{p}(t), \dot{x}(t)).$$

From (1.14) and (1.15), we may invert these relations to get

$$\dot{x}=\frac{\partial H}{\partial p}(x, p), \qquad -\dot{p}=\frac{\partial H}{\partial x}.$$

The advantage of considering the functional J instead of K is that, under reasonable conditions on H, it satisfies the Palais–Smale condition:

LEMMA 1.10. *Let H be strictly convex and differentiable, and let it satisfy the growth conditions:*

$$H(\lambda x, \lambda p) \geqq \lambda^\alpha H(x, p)>0$$

for some $\alpha>2$, all $\lambda>1$ and $(x, p)\neq 0$; and let

$$H(x, p) \leqq \frac{a}{2}(x^2+p^2)^{\alpha/2}$$

for all x, p and some $a>0$. Then J satisfies the Palais–Smale condition on the Banach space of $\dot{y}$, $\dot{q}\in L^{\alpha'}$ satisfying $\int_0^1 \dot{y}\, dt=\int_0^1 \dot{q}\, dt=0$.

The proof of Lemma 1.10 is relatively simple, and it is then an easy task to show that J satisfies the conditions of the mountain pass lemma. Thus for every $T>0$ one obtains a T-periodic solution of (1.12). Ekeland [29] has also shown that the energy of the solution satisfies the inequality

$$0<H(x(t), p(t))=h \leqq \frac{C}{T(\alpha/(\alpha-2))},$$

where C depends only on α and the minimum of H on the unit sphere.

Periodic solutions of Hamiltonian systems have also been obtained, by different methods, by A. Weinstein [63].

1.4. Variational methods for free boundary value problems. Variational inequalities have been used in several instances to treat free boundary value problems, but in some cases such problems may also be resolved by what are essentially isoperimetric problems. This has been done by Fraenkel and Berger for steady vortex rings [32] and by Temam [61], [62] for equilibrium configurations in a plasma.

The problem of steady vortex rings in an inviscid fluid of constant density is formulated as follows. Let r, θ, z be cylindrical coordinates and consider an axisymmetric flow with velocity $\vec{v}$ and vorticity $\vec{\omega}=\operatorname{curl}\vec{v}$. Since the density is constant, $\operatorname{div}\vec{v}=0$, so $\vec{v}$ may be written $\vec{v}=\operatorname{curl}\vec{A}$ where $\vec{A}$ is the vector potential. For an axisymmetric flow with no v_θ component, we may take $\vec{A}=(\psi/r)\hat{e}_\theta$ where ψ is the Stokes stream function. Then

$$\vec{v}=\left(-\frac{\psi_z}{r}, 0, \frac{\psi_r}{r}\right)=\operatorname{curl}\frac{\psi}{r}\hat{e}_\theta=\frac{1}{r}\nabla\psi\times\hat{e}_\theta$$

and

$$\vec{\omega} = \operatorname{curl} \vec{v} = -\frac{1}{r}(L\psi)\hat{e}_\theta$$

where

$$L = r\frac{\partial}{\partial r}\left(\frac{1}{r}\frac{\partial}{\partial r}\right) + \frac{\partial^2}{\partial z^2}.$$

Steady vortex flow is obtained when $\vec{\omega} = \omega\hat{e}_\theta$, and Euler's equations of steady axisymmetric flow require that ω/r be constant on stream lines. Therefore $\omega = \lambda r f(\psi)$ where f is called the vorticity function and λ is called the vorticity strength parameter. A vortex ring with cross section A in the plane $P = \{(r, z) \mid r > 0\}$ is obtained by setting $\omega = 0$ outside A and $\omega = \lambda r f(\psi)$ inside A. Equating this with the previous expression for ω we get the partial differential equation

$$L\psi = r\left(\frac{1}{r}\psi_r\right)_r + \psi_{zz} = \begin{cases} -\lambda r^2 f(\psi) & \text{in } A, \\ 0 & \text{in } P - A. \end{cases}$$

The boundary conditions on ψ are that ∂A and the z-axis be stream lines, and that the vortex ring move relative to the fluid at ∞ with velocity $(0, 0, W)$ (see Fig. 1.5). This implies the conditions

$$\psi|_{\partial A} = 0, \quad \psi|_{r=0} = -k \leqq 0, \quad \frac{\psi_z}{r} \to 0, \quad \frac{\psi_r}{r} \to -W$$

as $r^2 + z^2 \to \infty$.

In general one does not expect that $f(0+) = 0$, but only that $f(0+) > 0$. If we extend f to negative values by setting it equal to zero there, we may write our partial differential equation as

$$L\psi + \lambda r^2 f(\psi) = 0, \tag{1.16}$$

where f has a possible jump discontinuity at $\psi = 0$. f is assumed to be nondecreasing and positive for $\psi > 0$.

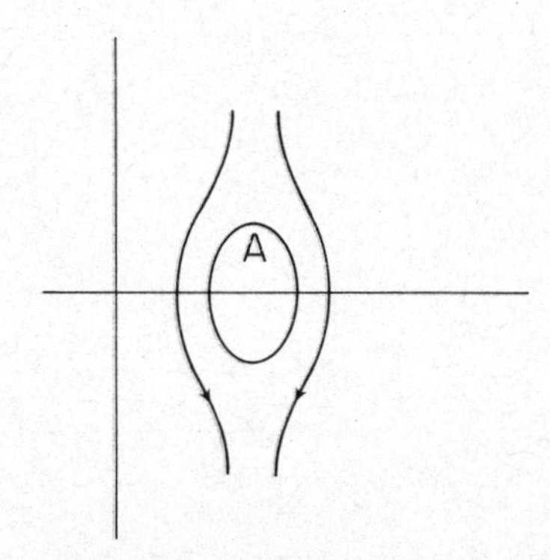

FIG. 1.5

The set $\{\psi=0\}$ is a free boundary for the problem. The two major analytical difficulties of the problem are the unbounded domain and the jump discontinuity of f. These are both overcome by limiting procedures, the first by approximating P by bounded domains $D=\{(r,z)\mid 0\leqq r\leqq r_0, |z|\leqq b\}$ and the second by approximating f by smooth functions. The boundary conditions on D are given by

$$\psi|_{\partial D}=-\tfrac{1}{2}Wr^2|_{\partial D}-k.$$

Let us write $\Psi=\psi-\frac{1}{2}Wr^2-k$ so that $\Psi|_{\partial D}=0$. Note that $Lr^2=0$ so that $L\Psi=L\psi$. The problem can now be given the following variational formulation: Given constants $W>0$ and $k>0$ and the kinetic energy η, find stationary points of the functional

$$J(\Psi)=\iint_D F(\Psi)r\,dr\,dz$$

on the sphere

$$\|\Psi\|^2=\iint_D \frac{1}{r^2}(\Psi_r^2+\Psi_z^2)r\,dr\,dz=\eta.$$

Here $F(u)=\int_0^u f(t)\,dt$.

One chooses a sequence of Ψ_n which maximizes J. Let $H(D)$ be the closure of $C_0^\infty(D)$ in the norm $\|\cdot\|$, and let $L_p(D,\tau)$ be the L_p space with measure $\tau=r\,dr\,dz$. Since

$$\iint(\nabla\psi)^2\,dr\,dz\leqq r_0\iint\frac{1}{r^2}|\nabla\psi|^2 r\,dr\,dz$$

and

$$\iint|\psi|^p\,r\,dr\,dz\leqq r_0\iint|\psi|^p\,dr\,dz,$$

we have the embeddings $H(D)\subset\mathring{W}_{1,2}(D)\subset L_p(D)\subset L_p(D,\tau)$, where the second embedding is compact. Therefore bounded sets in $H(D)$ are compact in $L_p(D,\tau)$ for every $p\geqq 1$. The constant λ appearing in (1.16) is a Lagrange multiplier and is given by

$$\lambda=\frac{\eta}{\iint\psi f(\Psi)r\,dr\,dz}.$$

The variational equations are

$$\langle\varphi,\psi\rangle=\lambda\iint_D\varphi f(\Psi)r\,dr\,dz$$

for all φ in $H(D)$; $\langle\cdot,\cdot\rangle$ is the quadratic form associated with the norm $\|\cdot\|$. Since $L(\frac{1}{2}Wr^2+k)=0$, $\langle\varphi,\psi\rangle=\langle\varphi,\Psi\rangle$.

The variational argument demonstrates the existence of a weak solution of (1.16). If f is Hölder continuous (even at 0) with exponent μ, the solution belongs to $C^{2+\mu}(\bar{D})$, but if f has a jump discontinuity at 0 then $\psi^{-1}(0)$ has measure 0, $\psi \in C^{1+\mu}(\bar{D})$ for $\mu \in (0, 1)$, and $\psi \in C^{2+\mu}$ in $D \backslash \psi^{-1}(0)$.

Uniform estimates are obtained for the vortex ring $A = \{(r, z) \mid \psi > 0\}$ and for λ as $D \nearrow P$, and a limiting solution on the unbounded domain is obtained.

The variational characterization of the solution can be used to obtain some information about its structure. for example:

THEOREM 1.11. *If f is convex and C^1, then A is simply connected.*

Proof. We have $f'(0) = 0$ and $F'(t) > 0$ for $t > 0$. If A has two components E_1 and E_2, then J is not maximized by Ψ on $\|\Psi\|^2 \eta$. Let

$$u(\beta) = \Psi \cos \beta + \eta^{1/2} v \sin \beta$$

where $\langle v, \Psi \rangle = 0$ and $\|v\| = 1$. Then $\|u(\beta)\|^2 = \eta$ and $u(0) = \Psi$. Expanding $J(u(\beta))$ for small values of β we get

$$\begin{aligned} J(u(\beta)) - J(\Psi) &= \frac{1}{2} \beta^2 \left\{ \eta \iint (f'(\Psi) v^2 - \Psi f(\Psi)) r \, dr \, dz \right\} + O(\beta^2) \\ &= \frac{1}{2} \beta^2 \left\{ \eta \iint f'(\psi) v^2 r \, dr \, dz - \frac{1}{\lambda} \langle \Psi, \psi \rangle \right\} + O(\beta^2). \end{aligned}$$

But $\langle \Psi, \psi \rangle = \langle \Psi, \Psi \rangle = \eta$, so

$$J(u(\beta)) - J(\Psi) = \frac{1}{2} \beta^2 \eta \left\{ \iint f'(\psi) v^2 r \, dr \, dz - \frac{1}{\lambda} \right\} + O(\beta^2).$$

Now suppose that the set $\{\psi > 0\}$ has two components E_1 and E_2. Choose $\varphi_j = \Psi = E_j$ and $\varphi_j = 0$ outside E_j. Then $\langle \varphi_1, \varphi_2 \rangle = 0$ and

$$\begin{aligned} \|\varphi_j\|^2 = \langle \varphi_j, \psi \rangle = \langle \varphi_j, \psi \rangle &= \lambda \iint_D \varphi_j f(\Psi) r \, dr \, dz \\ &= \lambda \iint_{E_j} \Psi f(\Psi) r \, dr \, dz. \end{aligned}$$

Choose constants c_1 and c_2 such that $v = c_1 \varphi_1 - c_2 \varphi_2$ satisfies

$$\|v\|^2 = c_1^2 \|\varphi_1\|^2 + c_2^2 \|\varphi_2\|^2 = 1,$$
$$\langle v, \psi \rangle = c_1 \langle \varphi_1, \psi \rangle - c_2 \langle \varphi_2, \psi \rangle = 0.$$

Then

$$\iint f'(\Psi) v^2 = \sum_j c_j^2 \iint_{E_j} f'(\Psi) \varphi_j^2$$

and

$$\frac{1}{\lambda} = \frac{1}{\lambda} \sum_j c_j^2 \langle \varphi_j, \varphi_j \rangle = \frac{1}{\lambda} \sum_j c_j^2 \iint_{E_j} \Psi f(\Psi),$$

so

$$\iint f'(\Psi) v^2 - \frac{1}{\lambda} = \sum_j c_j^2 \iint_{E_j} f'(\Psi) \Psi^2 - \Psi f(\Psi) > 0.$$

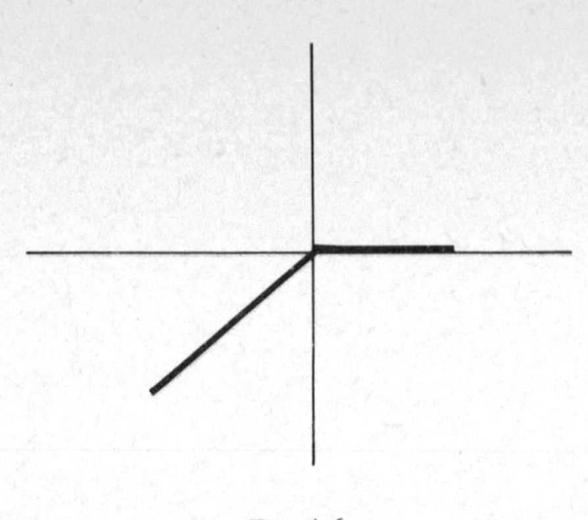

FIG. 1.6

Therefore the second variation of J at Ψ is strictly positive and Ψ does not maximize J if the set $\{\Psi>0\}$ has two components.

Temam (see also [13]) also considers isoperimetric problems arising from free boundary value problems. His boundary value problem is (essentially)

$$\Delta u+\lambda f(u)=0 \quad \text{in } \Omega,$$

$$\int_\Gamma \frac{\partial u}{\partial \nu}\,dx=I,$$

$$u|_\Gamma=\gamma=\text{const.}, \qquad \Gamma=\partial\Omega$$

where λ and γ are unspecified constants. The graph of the function f is shown in Fig. 1.6. Solutions of this boundary value problem are obtained as critical points of the functional

$$k_1(u)=\frac{1}{2}\iint(\nabla u)^2\,dx-Iv(\Gamma)$$

on the set

$$k_2(u)=\iint F(u)\,dx=\text{const.}$$

in the space $W=\{u \mid u\in H^1(\Omega),\ u=\text{const. on }\Gamma\}$.

1.5. Perturbation methods in critical point theory. A significant extension of the results on critical points of nonlinear functionals has been developed by A. Bahri and H. Berestycki [2]–[6]. They have developed these methods in conjunction with a classical topological approach to critical point theory for functionals invariant under a group action. Let us begin by considering the Hamiltonian system

$$\dot z=\mathcal{J}\nabla H(z)+f(t), \tag{1.17}$$

where $z=(q_1,\ldots,q_N,p_1,\ldots,p_N)\in\mathbb{R}^{2N}$, $\mathcal{J}=\begin{pmatrix}0 & I\\ -I & 0\end{pmatrix}$ and f is a T-periodic mapping from $\mathbb{R}$ to $\mathbb{R}^{2N}$. The Hamiltonian H is required to satisfy

$$\begin{aligned}&\text{a) } H\in C^2(\mathbb{R}^{2N},\mathbb{R}),\\ &\text{b) } H(z)\leqq \theta\langle\nabla H(z), z\rangle + C, \quad \text{where} \quad 0<\theta<\tfrac{1}{2} \text{ and } C>0,\\ &\text{c) } \frac{a}{p+1}|z|^{p+1}-b\leqq H(z)\leqq \frac{a'}{q+1}|z|^{q+1}+b',\end{aligned} \tag{1.18}$$

where $1<p\leqq q<2p+1$, $a', a>0$, $b', b\geqq 0$.

THEOREM 1.12 (Bahri and Berestycki [3], [5]). *Under the assumptions* (1.18 a, b, c) *the Hamiltonian system* (1.17) *possesses infinitely many T-periodic solutions.*

Rabinowitz [54] had previously shown that systems of the type $\dot{z} = \mathcal{J}\nabla \hat{H}_z(t, z)$ possess at least one T-periodic solution if $\hat{H}(t, z)$ is a T-periodic bounded perturbation of a super quadratic Hamiltonian $H(z)$.

Bahri and Berestycki have derived similar results for the inhomogeneous boundary value problem

$$\begin{aligned}-\Delta u = |u|^{p-1}u + h(x), \qquad x\in\Omega\subset\mathbb{R}^n,\\ u|_{\partial\Omega}=0, \qquad 1<p<\frac{N+2}{N-2}.\end{aligned} \tag{1.19}$$

In order to obtain their results, they applied a topological perturbation method developed by Bahri [2]. Consider the perturbed functional

$$I_f(u) = \int_\Omega |u|^{p+1} + f(x)u\, dx$$

on the sphere $\|u\|=1$ ($\|u\|$ is the Dirichlet norm of u). When $h=0$, the functional

$$I_0(u) = \int_\Omega \frac{|u|^{p+1}}{p+1}\, dx$$

is invariant under the group action $u\to -u$. The Euler–Lagrange equations for the critical points of I_0 on the sphere $\|u\|=1$ are $-\Delta u = \lambda\, |u|^{p-1}u$, where λ is the Lagrange multiplier. (However, λ can be eliminated by rescaling u in the homogeneous case.)

While the classical approach to obtaining critical points of I_0 is based on the notion of category, Bahri's methods depend on a different topological notion, that of genus. The functionals $I_0(u)$ and $\|u\|$ are invariant under the group action $u\to -u$ on the Banach space $E=\{u \mid \|u\|+\infty\}$. Let $\Sigma(E)$ denote the collection of compact symmetric (i.e. invariant under $u\to -u$) subsets $A\subset E$, with $0\notin A$. The *genus* of A, denoted by $\gamma(A)$, is defined by

$$\gamma(A) = \text{Min}\,\{k\in\mathbb{Z}^+ \mid \text{there exists an odd continuous map } \varphi: A\to S^{k-1}\}.$$

Here S^{k-1} denotes the $(k-1)$-dimensional sphere in $\mathbb{R}^k$. Note that an odd map in the present context is one which is equivariant with respect to the group action: thus $\varphi(-u) = -\varphi(u)$. A well-known theorem in topology, Borsuk's theorem, states that there is no odd continuous map from S^{k-1} to S^{j-1} for $j<k$. On the other hand, the identity map takes S^{k-1} into S^{j-1} for $j \geqq k$. Therefore, by Borsuk's theorem, $\gamma(S^{k-1})=k$.

THEOREM 1.13 (Ljusternik–Schnirelmann). *Let $f \in C^1(\mathbb{R}^N, \mathbb{R})$ be even and let S_r be a sphere of radius r in $\mathbb{R}^N$ whose center is at the origin. Then f has at least N distinct pairs of critical points. These are given by the minimax problems*

$$c_k = \underset{A \in \Gamma_k}{\text{Min}}\ \underset{x \in A}{\text{Max}}\, f(x),$$

where $\Gamma_k = \{A \mid A \in \Sigma(S_r),\ \gamma(A) \geqq k\}$.

Here $\Sigma(S_r)$ denotes the collection of all symmetric subsets of S_r.

The critical values c_k satisfy

1) $c_1 \leqq c_2 \leqq \cdots \leqq c_k \leqq c_{k+1}$.
2) $c_1 = \text{Min}_s f$, $c_N = \text{Max}_s f$.
3) If $c_j = c_{j+1} = \cdots = c_{j+p-1}$ then the set $K_c = \{x \mid f(x) = c,\ f'(x) = \lambda x\}$ has genus $\gamma(K_c) \geqq p$.

The classical Ljusternik–Schnirelmann proof of this result was based on a different topological notion, that of category. The proof that c_k is a critical value rests on the deformation lemma and proceeds in much the same way as the proofs of Theorems 1.1 and 1.2; one simply constructs equivalent deformations.

This classical result can be extended to functionals on Banach spaces which satisfy the Palais–Smale condition. For example, the functional $I_0(u)$ on the space $E = \{u \mid \|u\| < +\infty\}$ satisfies the Palais–Smale condition if $p < (n+2)(n-2)$. (The Palais–Smale condition is then a consequence of the Sobolev embedding theorem.) The problem (1.19) can thus be shown to have infinitely many pairs of distinct critical points. Coffman [24] and Hempel [39] derived such results using the notion of genus which we defined above. Browder [16], [17] had earlier given a very general development of Ljusternik–Schnirelmann theory for critical points of functionals which are invariant under a group of transformations with no fixed points. In addition to extending the theory to more general group actions, Browder also established the theory under minimal regularity assumptions on the functionals in order that the machinery be applicable to a wide class of nonlinear elliptic boundary value problems.

In the case of 2π-periodic solutions, one introduces the functional

$$I_f(z) = \int_0^{2\pi} \left\{ \frac{1}{2} \langle \dot{z}, \mathcal{J} z \rangle - \langle f, z \rangle \right\} dt.$$

This functional is invariant under the action of the group S^1 given by $(\tau z)(t) = z(t+\tau)$. Let E be the Banach space $[H^{1/2}(S^1)]^{2N}$—that is, of 2π-periodic mappings $z: S^1 \to \mathbb{R}^{2N}$ all of whose components have finite $H^{1/2}$ norm. Let E_m be the subspace of E spanned by the exponentials $\{E^{ijt} \mid |j| \leqq m\}$. A point in E_m

can be represented by its Fourier coefficients

$$u = \sum_{|j| \leqq m} a_j e^{ijt} \leftrightarrow (a_{-m}, \ldots, a_m)$$

where $a_j \in C^N$. (We identify points $(p_1, \ldots, p_N, q_1, \ldots, q_N)$ in $\mathbb{R}^{2N}$ with the points $(p_1 + iq_1, \ldots, p_N + iq_N) \in C^N$). The group action of S^1 on E_m becomes

$$\tau(a_{-m}, \ldots, a_m) = (e^{-im\tau} a_m).$$

Now S^1 also acts on odd-dimensional spheres S^{2n-1} in the following way. Points in S^{2n-1} may be identified with points $\zeta = (\zeta_1, \ldots, \zeta_n) \in C^n$ where $\sum_{j=1}^n |\zeta_j|^2 = 1$. The group action is *free*, that is, it has no fixed points; the same was true of the action $u \to -u$ on unit spheres.

There is an analogue of Borsuk's theorem for this S^1 action on odd-dimensional spheres, namely [9]:

THEOREM 1.14. *Let j, k be integers, $1 \leqq j < k$. There exists no continuous mapping $h: S^{2k-1} \to S^{2j-1}$ which is equivariant with respect to the group action $e^{i\theta}\zeta = (e^{i\theta}\zeta_1, \ldots, e^{i\theta}\zeta_n)$.*

Bahri and Berestycki [4], [6] use this theorem in exact analogy with the classical notion of genus for the Z_2 action $u \to -u$ on unit spheres to prove an S^1 analogue of the classical Ljusternik–Schnirelmann theorem. Define a set of mappings

$$\mathcal{H}_k^m = \{h : S^{2Nm-2k-1} \to E^M \backslash \{0\};\ h \text{ continuous and equivariant}\},$$

and a class of sets

$$\mathcal{A}_k^m = \{A \subset E^M \backslash \{0\};\ A = h(S^{2Nm-2k-1}),\ h \in \mathcal{H}_k^m\}.$$

then for $k \leqq m-1$ the numbers

$$C_k^m = \sup_{A \in \mathcal{A}_k^m} \operatorname{Min}_{z \in A} I_0(z)$$

are critical values of the functional I_0 restricted to E_m, and there exists an integer k_0 such that for all $k \geqq k_0$ there exists a subsequence $\{C_k^m\}$ which converges, as $m \to \infty$, to a critical value c_k of I_0.

The extension to the perturbed case I_f, $f \neq 0$, is based on an analysis of the homotopy group of the level sets. Let $B_{f,\alpha} = \{z \in E : I_f(z) \geqq \alpha\}$ and $B_{f,\alpha}^m = B_\alpha \cap E_m$. Bahri and Berestycki prove:

PROPOSITION 1.15. *Suppose that for some $\varepsilon > 0$, $C_{k-1}^m + \varepsilon < C_k^m - \varepsilon$. Then for any $W \subset E_m \backslash \{0\}$ such that*

$$B_{0,C_{k-\varepsilon}^m}^m \subset W \subset B_{0,C_{k-1}+\varepsilon}^m,$$

there exists $x_0 \in W$ for which the homotopy group $\pi(W; x_0)$ is nontrivial.

On the other hand, they can prove

PROPOSITION 1.16. *Let H satisfy hypotheses* (1.18 a, b, c) *above and suppose I_f possesses no critical values in (b, ∞). Then there exist integers k_0 and m_0 depending on b such that for all $a > b$, all $m \geqq m_0$ and all R such that $k_0 \leqq k \leqq m-1$ the homotopy group $\pi(B_{f,\alpha}^m, x_0) = 0$ for any $x_0 \in B_{f,\alpha}^m$.*

Using these two propositions, Bahri and Berestycki are able to prove that the Hamiltonian system (1.17) possesses an infinite sequence of 2π periodic solutions $\{z_k\}$ where the z_k are critical points of I_f such that $I_f(z_k) \to +\infty$ as $k \to \infty$. Their results represent a remarkable improvement over the classical results in this direction, where the demonstration of even one time periodic solution was a difficult proposition.

CHAPTER 2

Spontaneous Symmetry Breaking

2.1. Equivariant equations. A *representation* of a group $\mathscr{G}$ on a linear vector space V is a homomorphism $g \to T_g$ of $\mathscr{G}$ into the linear invertible operators on V; that is, $T_{g_1 g_2} = T_{g_1} T_{g_2}$ and $T_g^{-1} = T_g - 1$. A mapping G of V into itself is *covariant* (or sometimes *equivariant*) with respect to T_g if $T_g G(u) = G(T_g u)$ for all $g \in \mathscr{G}$.

This situation arises constantly in mathematical physics and is a mathematical formulation of the axiom that the equations of mathematical physics should be independent of the observer. Newtonian mechanics is characterized by invariance under the Euclidean group of rigid motions or one of its subgroups, while relativistic mechanics is characterized by invariance under the Poincaré group (the Lorentz group together with translations in space-time.)

Let $\mathscr{G}$ be a group of transformations acting on a manifold $\mathscr{M}$; we denote the action of $\mathscr{G}$ by $x \to x' = gx$ or by $x' = \varphi_g(x)$. Let Ψ be a scalar field defined on $\mathscr{M}$. We may suppose Ψ represents a physical quantity (for example temperature or density). If Ψ' is the scalar function for the same quantity computed in the new coordinate system, then the equivalence of the two representations requires

$$\Psi(x) = \Psi'(x'). \tag{2.1}$$

Let us denote the new scalar field Ψ' by $T_g \Psi$. Then (2.1) means $(T_g \Psi)(gx) = \Psi(x)$ or, replacing x by $g^{-1}x$,

$$(T_g \Psi)(x) = \Psi(g^{-1}x). \tag{2.2}$$

It is easily seen that (2.2) defines a representation T_g acting on the vector space $\mathscr{F}(m)$ of scalar functions on M. In this way, we see that the action of $\mathscr{G}$ on M induces, in a natural way, an action on the scalar fields $\mathscr{F}(M)$. It is clear that $\mathscr{F}(M)$ is linear and that the action T_g defined by (2.2) is thus a linear representation of $\mathscr{G}$ on the vector space $\mathscr{F}(M)$.

An example of an equivariant mapping is the Laplacian $\Delta = \sum_{i=1}^n \partial^2/\partial x^{i^2}$ acting on, say, $C^\infty(\mathbb{R}^n)$. The Laplacian is equivariant with respect to the group of rigid motions in $\mathbb{R}^n$.

The quantities of interest in physical theories are, however, generally not scalars, but more complicated objects such as tensors or spinors which reflect intrinsic properties of the system, and which transform under the group action according to more complicated transformation rules. For example, the Navier–Stokes equations for a viscous incompressible fluid are

$$\Delta u^i - \frac{\partial p}{\partial x^i} = u^j \frac{\partial u^i}{\partial x^j}, \qquad \frac{\partial u^i}{\partial x^i} = 0.$$

The physical quantity is $w=(\vec{u}, p)$ where p is the hydrodynamic pressure (a scalar field) and $\vec{u}$ is the velocity field (a vector field).

The action of a group $\mathcal{G}$ on a manifold m induces in a natural way an action of the group on the tangent bundle $T^1(M)$. Let us compute this action. Let $\varphi_g(x)$ be the diffeomorphism of M into itself corresponding to the group element g. Then φ_g satisfies

$$\varphi_{g_1g_2}(x)=\varphi_{g_1}(\varphi_{g_2}(x)). \tag{2.3}$$

Let $v(x)$ be a tangent vector at the point $x \in M$. To see what happens to $v(x)$, let $\gamma\,|t|$ be a curve in M such that $\gamma(0)=x$ and $\dot{\gamma}(0)=v(x)$. Then $\gamma(t)$ is carried under φ_g into the curve $\varphi_g(\gamma(t))$. The tangent vector to $\varphi_g(\gamma(t))$ at $t=0$ is

$$\frac{d}{dt}\varphi_g(\gamma(t))\big|_{t=0}=\varphi'_g\dot{\gamma}(0)=\varphi'_g v(x),$$

where φ'_g denotes the Jacobian of the transformation φ_g. Thus the action on $T^1(M)$ is given by

$$(x, v(x)) \to (\varphi_g(x), \varphi'_g(x)v(x)). \tag{2.4}$$

The vector which sits over $y=\varphi_g(x)$ is therefore $\varphi'_g(x)v(x)$; accordingly, the vector in the transformed field which sits over x is

$$(T_g v)(x)=(\varphi'_g(\varphi_g-1)(x))v(\varphi_g^{-1}x). \tag{2.5}$$

From the composition property (2.3) it follows that

$$\varphi'_{g_1g_2}(x)=\varphi'_{g_1}(\varphi_{g_2}(x))\varphi'_{g_2}(x) \tag{2.6}$$

and using the identity (2.6) one can show that the action (2.5) is a linear representation on vector fields; that is, $(T_{g_1g_2}v)(x)=(T_{g_1}(T_{g_2}v))(x)$.

For example, in the case of rigid motions on $\mathbb{R}^n$, a group element g takes the form $g=\{O, a\}$, and $gx=Ox+a$, where O is an orthogonal matrix and a is a vector in $\mathbb{R}^n$. The Jacobian of φ_g is therefore $\varphi'_g=0$, and the natural action (2.5) on vector fields on $\mathbb{R}^n$ is given by

$$(T_g v)(x)=(Ov)(g^{-1}x).$$

The Navier–Stokes equations for a viscous incompressible fluid are in fact equivariant with respect to the linear representation

$$T_g\begin{pmatrix}u^1\\u^2\\u^3\\p\end{pmatrix}(x)=\left(\begin{array}{ccc|c} & & & 0\\ & O & & 0\\ & & & 0\\ \hline 0 & 0 & 0 & 1\end{array}\right)\begin{pmatrix}u^1\\u^2\\u^3\\p\end{pmatrix}(g^{-1}x).$$

2.2. Equivariant bifurcation equations. Now suppose we have a bifurcation problem $G(\lambda, u)=0$ which is equivariant with respect to some group representation: $T_gG(\lambda, u)=G(\lambda, T_gu)$. Suppose that (λ_c, u_c) is a bifurcation point and that $T_gu_c=u_c$ for all g in $\mathcal{G}$. By the Lyapunov–Schmidt procedure the bifurcation problem can be reduced to a finite system of equations, the branching equations $F(\lambda, v)=0$. We have ([45, p. 87]):

THEOREM 2.1. *Let* u_c *be invariant under the symmetry group* $\mathcal{G}$, *let* G_u *be a Fredholm map, and let G be a differentiable mapping from the complex Banach space* $\mathcal{E}$ *to* $\mathcal{F}$, *with* $\mathcal{E} \subseteq \mathcal{F}$. *Then the kernel* $\mathcal{N} = \ker G_u(\lambda_c, u_c)$ *is invariant under* T_g *for* $g \in \mathcal{G}$, *and the bifurcation equations* $F(\lambda, v)$ *are equivariant with respect to the finite dimensional representation* $\Gamma_g = T_g|_{\mathcal{N}}$.

The resulting bifurcating solutions may have a smaller symmetry group; that is, the symmetry group of the bifurcating solutions may be only a subgroup of $\mathcal{G}$. In that case we have an example of *spontaneous symmetry breaking*. The symmetry group of the equations remains unchanged, but the solutions which bifurcate spontaneously break symmetry (in the absence of an external symmetry breaking perturbation).

The best known example of this situation is the Hopf bifurcation theorem in which temporal symmetry is broken. The autonomous equations $\dot{x} = F(\lambda, x)$ are invariant under the group of time translations $t \to t + \gamma$, as in an equilibrium (time independent) solution. The bifurcating time periodic solutions, however, are invariant only under the discrete subgroup generated by $t \to t + T$, T being the minimal period of the oscillations.

A very dramatic physical example of spontaneous symmetry breaking is the appearance of hexagonal convection cells or rolls in the Bénard experiments on convection in a thin layer of fluid heated from below. In this case Euclidean symmetry is broken to a crystallographic subgroup of $\mathcal{E}(2)$. The same type of symmetry breaking occurs in phase transitions, and in Ermentrout and Cowan's theory of hallucination patterns formed in the visual cortex [29]. For a discussion of physical applications of symmetry breaking to these and other areas, I refer the reader to the review article [57].

A second group that occurs commonly in physical applications is the rotation group. The breaking of rotational symmetry occurs in the buckling of spherical shells, the onset of convection in a spherical shell and in some morphogenesis processes [57]. The general problem of bifurcation in the presence of the rotation group is far from solved, though partial results have been obtained by Busse, Chossat, Golubitsky and Schaeffer, and myself. I shall present some new results for these problems in Chapter 4.

The branching equations assume the general structure

$$F(\lambda, v) = A(\lambda)v + B_2(\lambda, v; v) + B_3(\lambda, v, v) + \cdots, \tag{2.7}$$

where $A(\lambda)$ is a matrix, $B_\lambda(v, w)$ is a symmetric bilinear equivariant mapping, etc. By scaling the variables λ, v as follows:

$$\lambda = \varepsilon\tau, \qquad v = \varepsilon\xi,$$

we get

$$F(\varepsilon\tau, \varepsilon\xi) = \varepsilon^2(\sigma'(0)\xi + B_0(\xi, \xi)) + O(\varepsilon^2).$$

Dividing by ε^2 and letting $\varepsilon \to 0$ we arrive at the *reduced bifurcation equations*

$$\sigma'(0)\xi + B_0(\xi, \xi) = 0. \tag{2.8}$$

Such systems of equations (i.e. the linear and quadratic, or linear and cubic terms) have been considered by Busse, for the Bénard problem in planar and spherical geometrics. One must question the relationship of solutions of (2.8) to those of the original problem. The Jacobian of the mapping $\xi \to \sigma'(0)\xi + B_0(\xi, \xi)$ at a solution ξ_0 is $Jw = \sigma'(0)w + 2B_0(\xi_0, w)$. If this Jacobian is invertible it can be proved by the implicit function theorem that a solution of the full bifurcation equations (2.7) exists which is obtained as a perturbation of the solution of (2.8). The presence of a continuous transformation group, however, invalidates this approach. Suppose F is a mapping, equivariant with respect to a Lie group action T, and that $F(x_0) = 0$. Differentiating the identity $T_g F(x_0) = F(T_g x_0)$ with respect to the group parameters and setting $g = e$, we get $L_i F(x_0) = 0 = F'(x_0) L_i x_0$, where L_i is the Lie derivative with respect to the ith parameter and $F'(x_0)$ is the Jacobian matrix of F at x_0. Therefore $L_i x_0$ are null vectors of $F'(x_0)$, and $F'(x_0)$ is in general singular.

The upshot of this is that one cannot necessarily construct all solutions of the full bifurcation equations (1.7) as perturbations of the solutions of the reduced bifurcation equations (1.8). The implicit function theorem is too coarse a tool here, and a more refined perturbation approach is called for. Such investigations have been initiated by Golubitsky and Schaeffer [36], [37] and by Buzano and Golubitsky [20] in their applications of singularity theory to bifurcation problems. (See also E. N. Dancer [25], [26].)

The point is especially well illustrated by the recent work of Buzano and Golubitsky on the Bénard problem. In that case, the initial problem is equivariant with respect to the group of rigid motions in the plane, and one considers bifurcating doubly periodic solutions. The structure of the reduced bifurcation equations is derived in [56] on purely group theoretic grounds. By a certain "regularization technique" (see [56] for full details), solutions of the full problem may be obtained from solutions of the reduced bifurcation equations by an application of the implicit function theorem.

Using techniques of singularity theory, however, Buzano and Golubitsky show in the case where the leading nonlinear term is cubic that there are solutions of the full bifurcation equations which do not come from solutions of the reduced bifurcation equations. Such solutions vanish to a higher order than those which come from the reduced equations and arise due to terms of order 4 and 5. Buzano and Golubitsky's computations are nothing short of horrendous and I will leave the interested reader to pursue the details on his own. In the chapter on equivariant singularity theory I will carry out Golubitsky and Schaeffer's computations for the unfolding of the branch point for the representation $D^{(2)}$ of the rotation group.

2.3. Modules of equivariant mappings. As we observed in the previous section, the bifurcation equations are equivariant with respect to a finite dimensional representation of the symmetry group $\mathcal{G}$ of the problem. If $F(\lambda, v)$ is analytic in v (as it will be if the original functional equation $G(\lambda, u)$ is analytic) then it can be expanded in a power series of multilinear operators in v

(see (2.7)). The matrix $A(\lambda)$ intertwines with the representation; that is, $\Gamma_g A = A\Gamma_g$. If Γ_g is irreducible then by Schur's lemma A is a scalar multiple of the identity.

A more powerful computational approach arises from regarding the equivariant mappings as a module over the ring of invariant functions. Let Γ_g be a given representation acting on a vector space V. An *invariant* function on V is a real or complex valued function f for which $f(\Gamma_g x) = f(x)$ for all $x \in V$ and all $g \in \mathcal{G}$. The invariant functions form a commutative ring with identity. (A ring K is a set together with two operations, addition and multiplication, such that a) K is a commutative group with respect to addition; b) multiplication is associative; and c) $f(g+h) = fg + fh$ and $(f+g)h = fh + gh$. K is commutative if multiplication is commutative. The function $f \equiv 1$ is an identity for the ring of invariant functions.)

A *module* is roughly speaking a vector space over a commutative ring with identity. More precisely, a module M is a set of vectors which forms a commutative group with respect to addition and on which there is a "scalar multiplication" $(f, G) \to fG$ where G is a vector and f is an element of the commutative ring. (See Hoffmann and Kunze [41, p. 164].)

It is clear that the equivariant mappings constitute a module over the ring of invariant functions. We denote these respectively by $\mathscr{E}_\Gamma$ and I_Γ. (They depend on the representation Γ.)

A *Hilbert* basis for a ring R of invariant polynomials is a set of invariant polynomials $\sigma_1, \ldots, \sigma_n$ such that every $h \in R$ is a polynomial in $\sigma_1, \ldots, \sigma_n$. A module M is said to be finitely generated if there exist $G_1, \ldots, G_R$ in M such that every G in M can be expanded as a linear combination

$$G = f_1 G_1 + \cdots + f_k G_k$$

with coefficients $f_i \in R$. The rank of a finitely generated module is the smallest number of generators. The module is said to be free if the generators are linearly independent, that is, if $\sum_i f_i G_i = 0$ implies that all $f_i = 0$. Since elements of a ring do not have multiplicative inverses, it is possible that a set of generators of M is not free, but that none of the generators can be replaced by a linear combination of the others.

As an example, let the dihedral group D_n act on $\mathbb{R}^2$ by its two-dimensional irreducible representation. In complex coordinates z, $\bar{z}$ for $\mathbb{R}^2$ this action is generated by

$$z \to \bar{z} \quad \text{and} \quad z \to ze^{i\alpha}$$

where $\alpha = 2\pi/n$. We shall prove in Chapter 3 that a Hilbert basis for I_Γ is $u = z\bar{z}$ and $v = \operatorname{Re} z^n$, while a basis for the module $\mathscr{E}_\Gamma$ is given by z and $\bar{z}^{n-1}$. In these coordinates the general equivariant bifurcation equations therefore have the form

$$g(\lambda, z, \bar{z}) = a(\lambda, u, v)z + b(\lambda, u, v)\bar{z}^{n-1}. \tag{2.9}$$

Such a representation for the general structure of the bifurcation equations is obviously extremely convenient, but it depends on our being able to find all the invariants of the action and a basis for the module of equivariant mappings.

2.4. The rotation group. The problem of bifurcation in the presence of the rotation group leads to a number of interesting, and as yet unsolved, mathematical problems. The irreducible representations of the rotation group $SO(3)$ are denoted by D^l for $l=0, 1, 2, \ldots$ and are of dimension $2l+1$. For example, the spherical harmonics $Y^l_m(\theta, \varphi)=P_{l,m}(\cos\theta)e^{im\varphi}$, $-l \leqq m \leqq l$, transform according to a matrix representation of D^l of $SO(3)$. Let the kernel $\mathcal{N}$ of a rotationally invariant bifurcation problem transform according to D^l. Choosing a canonical basis $\Psi_{-l}, \ldots, \Psi_l$ for $\mathcal{N}$ (see [56]), we obtain branching equations in the form

$$F_m(\lambda, z_l, \ldots, z_l)=0, \qquad -l \leqq m \leqq l.$$

Since the representation is irreducible, the linear term of F_m is $\sigma(\lambda)z_m$ (see [56, p. 114]). If l is odd the quadratic terms vanish, while if l is even the quadratic terms take the form

$$\sum_{m_1+m_2=m} (-1)^m \begin{pmatrix} l & l & l \\ m_1 & m_2 & -m \end{pmatrix} z_{m_1} z_{m_2}$$

where $\begin{pmatrix} j_1 & j_2 & j_3 \\ m_1 & m_2 & m_3 \end{pmatrix}$ are the Wigner $3-j$ coupling coefficients for the rotation group.

The reduced bifurcation equations for l even are (up to a scale factor)

$$\lambda z_m = \sum_{m_1+m_2=m} (-1)^m \begin{pmatrix} l & l & l \\ m_1 & m_2 & -m \end{pmatrix} z_{m_1} z_{m_2}. \tag{2.10}$$

Busse [18] considered these equations for even l (although with respect to a different basis for $\mathcal{N}$) and constructed special solutions for all even l. The problem of finding all solutions of (2.10) and their symmetry groups for general l is completely open. Beyond that there is the question of actually obtaining all the solutions of the full branching equations.

A complete analysis of the reduced bifurcation equations for the case $l=2$ can be given in a concise form by making use of a special presentation of the representation $D^{(2)}$. Let O be an orthogonal matrix and consider the representation of $so(3)$ given by $T_0A=OAO^+$ acting on the nine-dimensional space M_3 of 3×3 real matrices. This representation is orthogonal relative to the inner product $(A, B)=\text{Tr}\, AB^t$. Now M_3 can be decomposed into the direct sum $M_3=V_2\oplus V_1\oplus V_0$ where V_2 is the five-dimensional space of symmetric matrices of trace zero, V_1 is the three-dimensional space of anti-symmetric matrices, and V_0 is the space λI, where I is the identity matrix. This orthogonal decomposition of M_3 corresponds to the Clebsch–Gordon series for T_0: namely, 0 is the representation D^1 of $so(3)$ and T_0 can be identified with $D^1\otimes(D^1)^t$. Since $(D^1)^t$ is equivalent to D^1, we have

$$T_0 \cong D^1\otimes D^1 = D^2\oplus D^1\oplus D^0.$$

Thus the representation OAO^t acting on V_2 contains a realization of the representation D^2. This presentation is very similar to the adjoint action of a Lie group on its Lie algebra. The adjoint action of $SU(3)$, for example, plays an important role in the theory of strong interactions in elementary particle physics. We shall discuss this action in §2.5.

Now the mappings $p_1(A)=A$ and $p_2(A)=A^2-\frac{1}{3}(\mathrm{Tr}\, A^2)I$ are equivariant with respect to the action T_0 and map $V_2 \to V_2$. It turns out (as we shall see in Chapter 3) that these are the only equivariant maps of orders one and two, and in fact they form a basis for the module of equivariant mappings. Thus the reduced bifurcation equations in this presentation are

$$A^2=\tfrac{1}{3}(\mathrm{Tr}\, A^2)I+\lambda A. \tag{2.11}$$

To solve (2.11) we simply note that every orbit under the action of T_0 passes through the diagonal matrices, so we may assume A is diagonal. Equations (2.11) are then readily solved.

The unfolding of this singularity, that is, the extension of solutions of (2.11) to the full branching equations, has been treated in thorough detail by Golubitsky and Schaeffer [37]. Their resolution of the problem was based on the fact, mentioned above, that each orbit of the action intersects the diagonal matrices. This allows one to reduce the problem to a branching problem on the two-dimensional subspace of traceless diagonal matrices. The symmetry group of this reduced problem is discrete. Let $H(\lambda, A)$ be equivariant with respect to the action T_0 on V_2 and let D be the subspace of diagonal matrices in V_2.

LEMMA 2.2 [37]. *D is invariant under H_1 and H is fully determined by its values on D.*

Proof. Since $OH(A)O^+=H(OAO^+)$ and since A is symmetric, we may choose D so that OAO^+ is diagonal. Then $H(A)=O^+H(OAO^+)O$ is determined. Furthermore, we see that O fixes $H(A)$ whenever O fixes A. It is easily seen that the subspace D may be characterized as the set of all A such that $O_iAO_i^t=A$ for

$$O_1=\begin{pmatrix}1 & 0 & 0\\ 0 & -1 & 0\\ 0 & 0 & -1\end{pmatrix}, \qquad O_2=\begin{pmatrix}-1 & 0 & 0\\ 0 & -1 & 0\\ 0 & 0 & 1\end{pmatrix}.$$

If $A\in D$ then A, hence $H(A)$, is fixed under O_i, and consequently $H(A)\in D$. Thus Lemma 2.2 is proved.

The subgroup of $O(3)$ which leaves D invariant is the group of 3×3 permutation matrices, and this group action on D is the two-dimensional representation of S_3. In fact, let $\mu=e^{2\pi i/3}$ and let $\mathrm{diag}\,(d_1, d_2, d_3)\in D$. We parametrize D by coordinates z, $\bar{z}$ as follows

$$\begin{pmatrix}d_1\\ d_2\\ d_3\end{pmatrix}=\begin{pmatrix}1 & 1\\ \mu & \mu^2\\ \mu^2 & \mu\end{pmatrix}\begin{pmatrix}z\\ \bar{z}\end{pmatrix}. \tag{2.12}$$

Then $d_1+d_2+d_3=0$ and d_1, d_2, d_3 are real. (Note that $\mu^2+\mu+1=0$.) The permutation (1 2 3) is effected by $(z, \bar{z}) \to (\mu z, \mu^2 \bar{z})$ and (2 3) is effected by $(z, \bar{z}) \to (\bar{z}, z)$. Thus the action of S_3 on D is generated by

$$z \to \mu z \quad \text{and} \quad z \to \bar{z}.$$

These actions generate a two-dimensional representation of S_3 which is, in fact, the two-dimensional representation of the dihedral group D_3. The canonical form of the equivariant mappings has already been stated above in (2.9), namely, in this case,

$$G(\lambda, z, \bar{z}) = a(u, v, \lambda)z + b(u, v, \lambda)\bar{z}^2,$$

where $u = z\bar{z}$ and $v = \operatorname{Re} z^3$. Given a general equivariant mapping $H: V_2 \to V_2$, we may therefore study its zero structure by restricting it to D and studying the zero structure of G. We shall return to this problem in Chapter 3 and again in Chapter 4.

2.5. *SU*(3) model for hadrons. The representation T_0 in §2.4 restricted to the subspace V_1 is the adjoint action of the Lie group $so(3)$ on its Lie algebra. The adjoint action of $SU(3)$ on its Lie algebra plays an important role in the theory of strong interactions of hadrons in elementary particle physics (see Gell-Mann and Ne'eman [33], also Michel and Radicati [45]).

The three-dimensional representation U, where $U \in SU(3)$, is denoted by 3 and its complex conjugate representation $\bar{U}$ is denoted by $\bar{3}$. U acts on a three-dimensional space V; we choose as a basis for V the vectors

$$u = \begin{pmatrix} 1 \\ 0 \\ 0 \end{pmatrix}, \quad d = \begin{pmatrix} 0 \\ 1 \\ 0 \end{pmatrix}, \quad s = \begin{pmatrix} 0 \\ 0 \\ 1 \end{pmatrix}.$$

The vectors u, d, s are interpreted physically as quarks which are called, respectively, up, down and strange. The adjoint transformation U^* acts on the dual space V^*; the basis vectors for V^* are taken to be

$$\bar{u} = (1, 0, 0), \quad \bar{d} = (0, 1, 0), \quad \bar{s} = (0, 0, 1),$$

and these vectors are interpreted physically as antiquarks. The representation obtained by taking the tensor product of 3 acting on V and $\bar{3}$ acting on the vector space V^* is denoted by $3 \otimes \bar{3}$. Let $\xi = (\xi^i) \in V$ and $\eta = \overline{\eta^i}) \in V^*$. Under a transformation U,

$$\xi^i \to U^{ij}\xi^j, \qquad \overline{\eta^i} \to \overline{\eta^j}\, \overline{U^{ij}},$$

so the tensor product $3 \otimes \bar{3}$, which acts on the tensor product $\xi^i \overline{\eta^j}$ as

$$\xi^i \overline{\eta^j} \to U^{ik}\xi^k \overline{\eta^l}\, \overline{U^{jl}} = \overline{U^{ik}}\, \xi^k \overline{\eta^l} (U^*)^{lj},$$

is equivalent to the adjoint action $A \to UAU^*$. This representation is unitary with respect to the inner product $\langle A, B \rangle = \operatorname{Tr} AB^*$, and the vector space W of Hermitian symmetric matrices, considered as a real nine-dimensional vector space, decomposes under this representation into $W = W_1 \otimes W_0$, where $W_0 = [I]$ and $W_1 = \{A \mid A = A^*, \operatorname{Tr} A = 0\}$. We discuss the orbits of $3 \otimes \bar{3}$ on W_1, the invariants and the critical points.

As a basis for the vector space W_1 we take the so-called λ-matrices of Gell-Mann:

$$\lambda_1 = \begin{pmatrix} 0 & 1 & 0 \\ 1 & 0 & 0 \\ 0 & 0 & 0 \end{pmatrix}, \quad \lambda_2 = \begin{pmatrix} 0 & -i & 0 \\ i & 0 & 0 \\ 0 & 0 & 0 \end{pmatrix}, \quad \lambda_3 = \begin{pmatrix} 1 & 0 & 0 \\ 0 & -1 & 0 \\ 0 & 0 & 0 \end{pmatrix},$$

$$\lambda_4 = \begin{pmatrix} 0 & 0 & 1 \\ 0 & 0 & 0 \\ 1 & 0 & 0 \end{pmatrix}, \quad \lambda_5 = \begin{pmatrix} 0 & 0 & -i \\ 0 & 0 & 0 \\ i & 0 & 0 \end{pmatrix}, \quad \lambda_6 = \begin{pmatrix} 0 & 0 & 0 \\ 0 & 0 & 1 \\ 0 & 1 & 0 \end{pmatrix},$$

$$\lambda_7 = \begin{pmatrix} 0 & 0 & 0 \\ 0 & 0 & -i \\ 0 & i & 0 \end{pmatrix} \quad \lambda_8 = \frac{1}{\sqrt{3}} \begin{pmatrix} 1 & 0 & 0 \\ 0 & 1 & 0 \\ 0 & 0 & -2 \end{pmatrix}.$$

The generators of the Lie algebra $su(3)$ may be obtained by multiplying each of these matrices by i. Thus the action $3 \otimes \bar{3}$ restricted to W_1 is equivalent to the adjoint action of $SU(3)$.

The Cartan subalgebra H of $su(3)$ is spanned by λ_3 and λ_8; it plays the role of the subspace D in the $SO(3)$ theory of §2.4. Since every $\lambda \in W_1$ can be diagonalized, every orbit of the adjoint action passes through the Cartan subalgebra. Let N_H be the normalizer subgroup of H—that is, $N_H = \{U \mid U \in SU(3),\ UHU^* \subseteq H\}$. The Cartan subalgebra generates an abelian subgroup of $SU(3)$, the maximal torus, which we denote by T_H. The *Weyl group* is the quotient group N_H/T_H, and in the present case the Weyl group is simply the permutation group S_3. For if $UHU^* \subseteq H$, then the action of U can only permute the diagonal entries, so that U must be a permutation matrix. (Similarly, for the general case $SU(n)$, the Weyl group is S_n.)

Since the Cartan subalgebra in the case of $SU(3)$ is two-dimensional, we are looking at the same two-dimensional representation of S_3 that we used in the $SO(3)$ theory. We may parametrize the Cartan subalgebra by coordinates z and $\bar{z}$ just as we did in (2.12), and the invariants of the adjoint action are $u = z\bar{z}$ and $v = \operatorname{Re} z^3$, just as before. The critical points of v on the surface $u = 1$ are pictured in Fig. 2.1. We have added the point at the origin, thereby obtaining the vector figure (root diagram) for the Lie algebra $su(3)$. In the $su(3)$ model for hadrons, each point on this figure, that is, every vector in the Lie algebra $su(3)$, corresponds to an elementary particle; these are labeled above. π^+ and π^- are the positively and negatively charged pions, and K^+, K^-, K^0 and $\overline{K^0}$ are kaons. The particles at the origin are π^0 and η^0; they are neutral and do not interact with either the electric or strong forces. The directions of charge and hypercharge of the particles are indicated at the right. Thus, for example, the pions $\pi^\pm$ have electric charges ± 1. The kaon $\overline{K^0}$ has a fractional electric charge and hypercharge $-\sqrt{3}/2$, etc. (See [34, p. 283].)

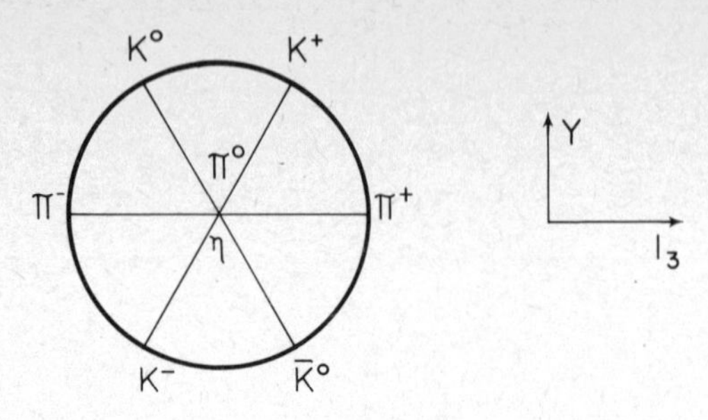

FIG. 2.1

Now recall that we obtained the vectors above as tensor products of the space V and V^*. In other words, the hadrons are regarded as composite particles made up of the more elementary quarks. Thus, for example

$$u\bar{d} \sim \begin{pmatrix} 0 & 1 & 0 \\ 0 & 0 & 0 \\ 0 & 0 & 0 \end{pmatrix} \quad \text{and} \quad d\bar{u} \sim \begin{pmatrix} 0 & 0 & 0 \\ 1 & 0 & 0 \\ 0 & 0 & 0 \end{pmatrix},$$

so

$$u\bar{d} + d\bar{u} \sim \begin{pmatrix} 0 & 1 & 0 \\ 1 & 0 & 0 \\ 0 & 0 & 0 \end{pmatrix}.$$

It may be worth noting the following generalization of the parametrization (2.12) for the real traceless diagonal matrices of order $(2m+1)\times(2m+1)$. Let $\lambda = e^{2\pi i/2m+1}$ and construct the matrix

$$\Omega = \begin{pmatrix} 1 & 1 & \cdots & 1 \\ \lambda & \lambda^2 & \cdots & \lambda^2 \\ \lambda^2 & \lambda^4 & \cdots & \\ \vdots & & & \\ \lambda^{2m} & \lambda^{2m-1} & \cdots & \lambda \end{pmatrix}.$$

We parametrize the diagonal matrices $\operatorname{diag}(d_1, \ldots, d_{2m+1})$ by the complex variables $z_1, z_2, \ldots, z_m$ by setting

$$\begin{pmatrix} d_1 \\ d_2 \\ \vdots \\ d_{2m+1} \end{pmatrix} = \Omega \begin{pmatrix} z_1 \\ z_2 \\ \vdots \\ z_m \\ \bar{z}_m \\ \vdots \\ \bar{z}_1 \end{pmatrix}.$$

Then $d_1, \ldots, d_{2m+1}$ are real and $d_1 + \cdots + d_{2m+1} = 0$. If P_σ is the representation of S_n by permutation matrices and T_σ is the corresponding representation on the variables $z_1, \ldots, z_m$, we have $d \to P_\sigma d$, $z \to T_\sigma z$, hence $\Omega T_\sigma = P_\sigma \Omega$. We can invert this and solve for T_σ. In fact,

$$\Omega^t \Omega = (2m+1)\Delta$$

where

$$\Delta' = \begin{pmatrix} 0 & & 1 \\ & \cdot^{\cdot^{\cdot}} & \\ 1 & & 0 \end{pmatrix}.$$

Therefore $\Delta^2 = 1$ and so $T_\sigma = (1/(2m+1))\Delta\Omega^t P_\sigma \Omega$. T_σ is thus an m-dimensional representation of S_n, and in fact is a representation of the Weyl group acting on the Cartan subalgebra. Higher dimensional Lie groups are of possible interest in unified field theories.

2.6. Hopf bifurcation in the presence of spatial symmetries. In this section we give a presentation of the Hopf bifurcation theorem when spatial symmetries are present. Such situations lead physically to the bifurcation of waves from a resting state. Traveling and standing waves on an infinite line have been discussed in connection with neurological nets by Ermentrout and Cowan [28], and rotating and standing waves in a circular geometry have been discussed by Herschkowitz-Kaufman and Erneux [40]. Another geometry of physical interest is the spherical geometry: what type of wave motion can be expected to appear on the surface of a sphere? We derive here the general structure of the bifurcation equations in a convenient form and in the next section present a streamlined stability analysis.

Consider the bifurcation of time periodic solutions of $u_t - G(\lambda, u) = 0$ from the resting state $u = 0$. Assuming $u = \lambda = 0$ is the bifurcation point, these equations may be linearized at the origin to obtain

$$u_t - L(\lambda)u + R(\lambda, u) = 0. \tag{2.13}$$

The solutions of (2.13) are assumed to take their values in a Banach space ε. We assume that a pair of complex conjugate eigenvalues of $L(\lambda)$ crosses the imaginary axis when λ crosses zero: $L(\lambda)\varphi(\lambda) = \gamma(\lambda)\varphi(\lambda)$, with $\gamma(0) = i\omega_0$ and $\operatorname{Re}\gamma'(0) > 0$. We assume further that (2.13) is equivariant with respect to a representation T_g of a compact group $\mathcal{G}$. Consequently the eigenspace $\mathcal{N}_i = \operatorname{Ker}(L(0) - i\omega_0 I)$ is invariant under a finite dimensional representation Γ_g of $\mathcal{G}$. We may choose units of time in which $\omega_0 = 1$. Setting $s = \omega t$, we obtain the following form for equations (2.13):

$$\omega \frac{\partial u}{\partial s} - L(\lambda)u + R(\lambda, u) = 0$$

where $u(s)$ is 2π-periodic in s. The details of the Lyapunov–Schmidt reduction method are given in [56]. The projection onto Ker J is given by

$$Pu = \sum_m \langle u, \Psi_m^* \rangle \Psi_m + \langle u, \overline{\Psi_m^*} \rangle \overline{\Psi_m}, \tag{2.14}$$

where $\{\Psi_m^*\}$ are the null functions of the adjoint operator J^*. Specifically $\Psi_m^* = e^{-is}\varphi_m^*$ where $L_0^*\varphi_m^* = i\varphi_m^*$. The bilinear form $\langle u, \Psi_m^* \rangle$ on the space of 2π-periodic functions with values in $\mathscr{E}$ is given by

$$\langle u, \Psi_m^* \rangle = \frac{1}{2\pi} \int_0^{2\pi} \langle u(s), \Psi^* \rangle_{\mathscr{E}} \, ds$$

where $\langle \cdot, \cdot \rangle_{\mathscr{E}}$ is the bilinear pairing between $\mathscr{E}$ and its dual. The vectors φ_m, φ_m^* are assumed to be normalized so that $\langle \varphi_m, \varphi_n^* \rangle_{\mathscr{E}} = \delta_{mn}$; then $\langle \Psi_m, \Psi_n^* \rangle = \delta_{mn}$.

Applying P and $Q = I - P$ to the time dependent equation we get the coupled system

$$\begin{aligned} QJu + (\omega - 1)Q\frac{\partial u}{\partial s} - Q(L(\lambda) - L_0)u + QR(\lambda, u) &= 0, \\ PJu + (\omega - 1)P\frac{\partial u}{\partial s} - P(L(\lambda) - L_0)u + PR(\lambda, u) &= 0. \end{aligned} \tag{2.15}$$

Putting $u = v + \chi$ where $v = Pu$, the first equation may be solved for $\chi = \chi(\lambda, \omega, v)$ as a function of v by the implicit function theorem. The bifurcation equations are then obtained from the second equation in (2.15) by substituting $u = v + \chi(\lambda, \omega, v)$. Since $Jv = 0$ and $PJ\chi = 0$ we get, to lowest order,

$$(\omega - 1)P\frac{\partial}{\partial s}\left(\sum_m z_m \Psi_m + \overline{z_m}\,\overline{\Psi_m}\right) - P(L(\lambda) - L_0)\left(\sum_m z_m \Psi_m + \overline{z_m}\,\overline{\Psi_m}\right) + \cdots = 0. \tag{2.16}$$

Now $\partial\Psi_m/\partial s = i\Psi_m$, and $\overline{\partial\Psi_m/\partial s} = -i\overline{\Psi_m}$. Moreover, we claim that $L(\lambda)\Psi_m = \gamma(\lambda)\Psi_m$ and $L(\lambda)\overline{\Psi_m} = \overline{\gamma(\lambda)}\,\overline{\Psi_m}$, at least if a certain condition is satisfied. We first establish:

THEOREM 2.3. *Let $\mathcal{N}_i$ be irreducible and let T_g contain the irreducible representation Γ only once. Then $L(\lambda)$ leaves $\mathcal{N}_i$ invariant and $L(\lambda)|_{\mathcal{N}_i} = \gamma(\lambda)I$.*

Proof. If T contains Γ only once, then there is a uniquely defined projection E onto $\mathcal{N}_i$ which commutes with T; in fact, E is given by

$$E = \frac{n_\nu}{|\mathscr{G}|} \int \overline{\chi^{(\nu)}(g)} T_g \, d\mu(g)$$

where $n_\nu = \dim \Gamma = \dim \mathcal{N}_i$, $|\mathscr{G}|$ is the volume of the compact group $\mathscr{G}$, $d\mu$ is the invariant measure, and $\chi^{(\nu)}(g)$ is the character of the representation Γ. Since T_g commutes with $L(\lambda)$, so does E, and therefore $L(\lambda)$ leaves $\mathcal{N}_i$ invariant. Since $\mathcal{N}_i$ is irreducible and $L(\lambda)$ commutes with Γ_g, it is a scalar multiple of the identity by Schur's theorem, and thus the conclusion of Theorem 2.3 follows.

It follows immediately that $L(\lambda)$ leaves Ker J invariant, and that $L(\lambda)\Psi_m = \gamma(\lambda)\Psi_m$. From the representation (2.14) we can show that P commutes with

$L(\lambda)$. In fact,

$$\begin{aligned} PL(\lambda)u &= \sum_m \langle L(\lambda)u, \Psi_m^*\rangle \Psi_m + \langle L(\lambda)u, \Psi_m^*\rangle \Psi_m \\ &= \sum_m \langle u, L^*(\lambda) e^{is} \varphi_m^*\rangle \Psi_m + \cdots \\ &= \sum_m \langle u, \gamma(\lambda) e^{is} \varphi_m^*\rangle \Psi_m + \cdots \\ &= \sum_m \langle u, \Psi_m^*\rangle \gamma(\lambda) \Psi_m + \cdots \\ &= L(\lambda) Pu. \end{aligned}$$

Returning to (2.16) we get

$$\sum_m [i(\omega-1)-(\gamma(\lambda)-\nu(0))] z_m \Psi_m + [-i(\omega-1)-(\bar{\gamma}(\lambda)-\bar{\gamma}(0))]\overline{z_m}\,\overline{\Psi_m} + \cdots = 0, \tag{2.17}$$

or, since $\gamma(0)=i$,

$$\sum_m (i\omega-\gamma(\lambda)) z_m \Psi_m + (i\omega - \bar{\gamma}(\lambda))\overline{z_m}\,\overline{\Psi_m} + \cdots = 0. \tag{2.18}$$

We have thus determined the lowest order terms of the bifurcation equations.

In order to get a concise form of the structure of the higher order terms in the bifurcation equations, we now look at them from the group theoretic point of view. The symmetry group for the time periodic solutions is $SO(1)\times\mathcal{G}$. Note that $t\to -t$ is not a symmetry of (2.13). The representation of $SO(1)\times\mathcal{G}$ on $\ker J$ is contained in the tensor product of the representations Γ_g and

$$\sigma(\theta) = \begin{pmatrix} e^{i\theta} & 0 \\ 0 & e^{-i\theta} \end{pmatrix}.$$

Elements in $\operatorname{Ker} J$ can be represented as $e^{is}\Psi + e^{-is}\varphi$, where $\Psi\in\mathcal{N}_i$ and $\varphi\in\eta_{-i}$. Since we require invariance under complex conjugation (real solutions), we take $\varphi=\bar{\Psi}$. The tensor product $\sigma\otimes\Gamma$ acts on such elements by

$$\sigma\otimes\Gamma(e^{is}\Psi + e^{-is}\bar{\Psi}) = e^{i\theta}(e^{is}\Gamma\Psi) + e^{-i\theta}(e^{-is}\Gamma\bar{\Psi}).$$

A mapping F on $\operatorname{Ker} J$ to itself has the general structure

$$F(e^{is}\Psi + e^{-is}\bar{\Psi}) = p(\Psi,\bar{\Psi})e^{is} + \overline{p(\Psi,\bar{\Psi})}e^{-is}.$$

The equivariance of F leads to the following condition on p:

$$e^{i\theta}\Gamma_g(p(\Psi,\bar{\Psi})) = p(e^{i\theta}\Gamma_g\Psi, e^{-i\theta}\Gamma_g\bar{\Psi}). \tag{2.19}$$

The branching equations are therefore

$$p(\Psi,\bar{\Psi},\lambda,\omega)=0, \qquad \overline{p(\Psi,\bar{\Psi},\lambda,\omega)}=0. \tag{2.20}$$

Both equations in (2.20) must be used in a stability analysis of the branching solutions, but, due to the special nature of (2.20), it is sufficient to solve only one:

$$p(\Psi, \bar{\Psi}, \lambda, \omega) = 0, \tag{2.21}$$

in order to obtain all bifurcating solutions. By comparison with (2.18) we see that

$$p(\Psi, \bar{\Psi}, \lambda, \omega) = (i\omega - \gamma(\lambda)) + q(\lambda, \omega, \Psi, \bar{\Psi})$$

where q transforms as in (2.19).

Let us consider some special cases. When no spatial symmetries are present, η_i is one-dimensional and we may take $\Psi = z\varphi_1$, where $[\varphi_1] = \eta_i$. Then (2.21) reduces to $p(z, \bar{z}, \lambda, \omega) = 0$ where p transforms according to (2.19)—that is,

$$e^{i\theta} p(z, \bar{z}, \lambda, \omega) = p(e^{i\theta} z, e^{-i\theta} \bar{z}, \lambda, \omega).$$

The only invariant of this action is $z\bar{z}$, and the module of equivariant mappings is of rank one and is spanned by the generator $F_n(z, \bar{z}) = (z, \bar{z})$. Therefore $p = za(|z|^2, \lambda, \omega)$. Letting $\sigma = |z|^2$, a takes the form[1]

$$a = i(\omega - 1) - \lambda\gamma'(0) + a_1\sigma + a_2\sigma^2 + \cdots \tag{2.22}$$

where the coefficients $a_1, a_2, \ldots$ depend on λ and ω and in general are complex. The bifurcation equations in this case (the classical Hopf bifurcation theorem) are therefore

$$a(\sigma, \lambda, \omega) = 0.$$

Now consider the case where equations (2.13) are posed on the real line $-\infty < x < \infty$ and are equivariant with respect to the symmetries

$$x \to x + \gamma, \qquad x \to -x.$$

This occurs, for example, when the original problem is defined on the real line and is invariant under translations and reflections, and we look for bifurcating solutions which are periodic in both space and time. A similar situation occurs when the problem is posed in a circular geometry and is invariant under the symmetries

$$O(2): \theta \to \theta + \gamma, \qquad \theta \to -\theta.$$

There is one difference between these two problems. In the case of circular symmetry the periodicity in θ is predetermined, whereas in the case of spatially periodic solutions the wave number is not determined.

Let $\mathcal{N}_i = \{\Psi_1, \Psi_2\}$ and assume the group of space translations $x \to x + \gamma$ acts in such a way that $T_\gamma \Psi_1 = e^{ik\gamma}\Psi_1$, $T_\gamma \Psi_2 = e^{-ik\gamma}\Psi_2$. For example, one might

---

[1] In the case where no symmetry is present, we can no longer make the assertion that $L(\lambda)$ leaves $\mathcal{N}_i$ invariant. Recall that $\mathcal{N}_i = \operatorname{Ker}(L_0 - i)$. In general this space will change with λ; that is, $\mathcal{N}_i \neq \operatorname{Ker}(L(\lambda) - i)$. In this case one can show by perturbation theory ([45, p. 53]) that $\langle L_1\varphi_0, \varphi_0^* \rangle = \gamma'(0)$ (we are writing $L(\lambda) = L_0 + \lambda L_1 + \cdots$). Therefore in this case $P(L(\lambda) - L_0)u = \gamma'(0)\lambda u + O(\lambda^2)$.

have $\Psi_1 = e^{ikx}\varphi$, $\Psi_2 = e^{-ikx}\varphi$ where φ is a vector in some Banach space. We denote the reflection $x \to -x$ by P. Then $P\Psi_1 = \Psi_2$ and $P\Psi_2 = \Psi_1$.

A general point Ψ in $\mathcal{N}_i$ can be represented by $\Psi = z_1\Psi_1 + z_2\Psi_2$. A set of coordinates on the real subspace of Ker J is thus given by $(z_1, z_2, \bar{z}_1, \bar{z}_2)$. We denote time translations as before by σ_θ. The operations σ_θ, T_γ and P then have the following matrix representation relative to this basis:

$$\sigma_\theta = \begin{pmatrix} e^{i\theta} & & & 0 \\ & e^{i\theta} & & \\ & & e^{-i\theta} & \\ 0 & & & e^{-i\theta} \end{pmatrix}, \quad T_\gamma = \begin{pmatrix} e^{ik\gamma} & & & \\ & e^{-ik\gamma} & & \\ & & e^{-ik\gamma} & \\ & & & e^{ik\gamma} \end{pmatrix},$$

$$P = \begin{pmatrix} 0 & 1 & 0 & 0 \\ 1 & 0 & 0 & 0 \\ 0 & 0 & 0 & 1 \\ 0 & 0 & 1 & 0 \end{pmatrix}. \tag{2.23}$$

The invariants of σ_θ are generated by $\{z_i\bar{z}_j\}$, $i, j = 1, 2$, and those of T_γ are generated by $|z_1|^2$, $|z_2|^2$, z_1z_2 and $\bar{z}_1\bar{z}_2$. The invariants of both are generated by $r = |z_1|^2$ and $s = |z_2|^2$. When we require in addition that the quantities be invariant under P, we see that $\sigma = r + s$ and $\tau = rs$ constitute a Hilbert basis for the symmetries (2.23).

Now we construct a basis of generators for the module of equivariant mappings. The general mapping is of the form $F = (F_1, F_2, \bar{F}_1, \bar{F}_2)$. The equivariance conditions on F_1 and F_2 are

$$\begin{aligned} \sigma_\theta:&\quad e^{i\theta}F_1(z_1, z_2, \bar{z}_1, \bar{z}_2) = F_1(e^{i\theta}z_1, e^{i\theta}z_2, e^{-i\theta}\bar{z}_1, e^{-i\theta}\bar{z}_2), \\ T_\gamma:&\quad e^{ik\gamma}F_1(z_1, z_2, \bar{z}_1, \bar{z}_2) = F_1(e^{ik\gamma}z_1, e^{-ik\gamma}z_2, e^{-ik\gamma}\bar{z}_1, e^{ik\gamma}\bar{z}_2), \\ P:&\quad F_2(z_1, z_2, \bar{z}_1, \bar{z}_2) = F_1(z_2, z_1, \bar{z}_2, \bar{z}_1). \end{aligned}$$

Expanding F_1 and F_2 in polynomials in $z_1, z_2, \bar{z}_1, \bar{z}_2$, we may consider each monomial separately. If $F_1 = z_1^a z_2^b \bar{z}_1^c \bar{z}_2^d$, then by the symmetry P, $F_2 = z_1^b z_2^a \bar{z}_1^d \bar{z}_2^c$. The symmetries σ_θ and T_γ lead to the equations

$$1 = a - b - c + d, \qquad 1 = a + b - c - d.$$

These equations are satisfied by taking $a = c + 1$ and $c = d$. Consequently the generators for F_1 are all of the form $F_1 = r^c s^b z_1$, and the module of mappings of C^2 into C^2, which are equivariant with respect to σ_θ, T_γ, and P, is generated by the mappings

$$\begin{pmatrix} r^c s^b z_1 \\ r^b s^c z_2 \end{pmatrix},$$

where b and c are integers. (I am assuming here that the branching equations are analytic. The extension to C^∞ mappings depends on theorems of Schwarz [59] and Poenaru [51]; see also Golubitsky and Schaeffer [36].) Now I claim that this module is actually of rank two and generated by

$$f_1=\begin{pmatrix} z_1 \\ z_2 \end{pmatrix} \quad \text{and} \quad f_2=\begin{pmatrix} rz_1 \\ sz_2 \end{pmatrix}.$$

This follows from induction on the two recursion formulae

$$\begin{pmatrix} r^{m+1}s^n z_1 \\ r^n s^{m+1} z \end{pmatrix}=\sigma\begin{pmatrix} r^m s^n z_1 \\ r^n s^m z_2 \end{pmatrix}-\tau\begin{pmatrix} r^{m-1}s^n z_1 \\ r^n s^{m-1} z_2 \end{pmatrix},$$

$$\begin{pmatrix} r^m s^{n+1} z_1 \\ r^n s^{m+1} z_2 \end{pmatrix}=\sigma\begin{pmatrix} r^m s^n z_1 \\ r^n s^m z_2 \end{pmatrix}-\tau\begin{pmatrix} r^m s^{n-1} z_1 \\ r^{n-1} s^m z_2 \end{pmatrix}.$$

Thus the first two components of the bifurcation mapping can be written in the form $F=af_1+bf_2$, where a, b are functions of λ, ω and the invariants σ and τ. The full four components are therefore

$$F=\begin{pmatrix} (a+br)z_1 \\ (a+bs)z_2 \\ (\bar{a}+\bar{b}r)\bar{z}_1 \\ (\bar{a}+\bar{b}s)\bar{z}_2 \end{pmatrix}.$$

The branching equations are

$$(a+br)z_1=0, \qquad (a+bs)z_2=0, \tag{2.24}$$

where, by comparison with (2.18),

$$a(\lambda, \omega, \sigma, \tau)=i\omega-\gamma(\lambda)+a_1\sigma+a_2\tau+\cdots.$$

There are three types of solutions that one may consider here:

(i) $r=s$, $a+br=0$,
(ii) $s=0$, $a+br=0$,
(iii) $a=0$, $b=0$.

The first case can be dealt with as follows. When $r=s$ we have $\sigma=2r$, $\tau=r^2$, and so the equation $a+br=0$ takes the form

$$i\omega-\gamma(\lambda)+rh(\lambda, \omega, r)=0.$$

Taking real and imaginary parts of this equation, we get

$$h_1=-\text{Re}\,\gamma(\lambda)+r\,\text{Re}\,h(\lambda, \omega, r)=0,$$
$$h_2=\omega-\text{Im}\,\gamma(\lambda)+r\,\text{Im}\,h(\lambda, \omega, r)=0.$$

these equations can be solved for $\lambda=\lambda(r)$ and $\omega=\omega(r)$ since the Jacobian at $\lambda=0$, $\omega=1$, $r=0$ is

$$\frac{\partial(h_1, h_2)}{\partial(\lambda, \omega)}=\begin{pmatrix} -\text{Re}\,\gamma'(0) & 0 \\ -\text{Im}\,\gamma'(0) & 1 \end{pmatrix},$$

which is invertible if we assume the Hopf transversality condition $\text{Re}\,\gamma'(0)\neq 0$.

The second case can be resolved in virtually the same way. When $s=0$, then $\sigma=r$ and $\tau=0$, and the equation $a+br=0$ is similar to the one obtained above.

The third case, $a=0$, $b=0$, is altogether different. Here we must solve $a=0$, $b=0$ simultaneously. Since the coefficients of a and b are in general complex, while λ, ω, r and s are real, this amounts to solving four (real) equations in four (real) unknowns.

In general we should not expect to be able to solve such a system, say for $\lambda=\lambda(r,s)$, $\omega=\omega(r,s)$. The situation may be compared to the problem of trying to resolve the Hopf bifurcation problem without allowing the frequency to vary. It one naively attempts to find bifurcating periodic solutions without treating the frequency as a parameter to be determined, one obtains the bifurcation equations (2.22) with $\omega=1$, namely,

$$a\equiv-\lambda\gamma'(0)+a_1\sigma+a_2\sigma^2+\cdots=0,$$

where the coefficients a_i are complex. (Note that if one had the symmetry $t\to-t$ one could show that the coefficients a_i are real, but the equations $\dot{x}=F(x,\lambda)$ are not invariant under time reversal. See [56, pp. 123–125].) Since the a_i are complex, the branching equation is equivalent to two real equations in the two real unknowns λ and σ, and is therefore in general unsolvable.

To return to the case at hand, let us try to introduce an additional parameter into the problem in a natural way. If we are dealing with the bifurcation of waves on the line, the natural parameter to introduce is the wave number k. Scaling both the space and time variables by setting $s=\omega t$ and $\xi=kx$, the solution takes the form $u(x,t)=v(kx,\omega t)$, where $v(\xi,s)$ is 2π-periodic in both ξ and s. Assuming that $L(\lambda)$ and $R(\lambda,\cdot)$ involve only spatial derivatives, we may write the equations in the form

$$\omega\frac{\partial v}{\partial s}-L(\lambda,k^2)v+R(\lambda,k^2,v)=0.$$

We indicate dependence on k^2 since we are assuming the equations are invariant under the symmetry $x\to-x$. If the critical eigenvalue is now denoted by $\gamma(\lambda,k^2)$, that is, $L(\lambda,k^2)\varphi=\gamma(\lambda,k^2)\varphi$, then, retracing the analysis that led to (2.18), we arrive this time at the following expression for the lowest order of a:

$$a(\lambda,w,k^2,\sigma,\tau)=i\omega-\gamma(\lambda,k^2)+a_1\sigma+\cdots.$$

The form of the leading terms of b is harder to determine, involving higher order perturbations in the Lyapunov–Schmidt procedure. The introduction of the wave number k, however, has led us to a system of four equations in five unknowns, of which the first two take the form

$$-\text{Re}\,\gamma(k^2,\lambda)+\text{Re}\,(a_1\sigma+a_2\tau+\cdots)=0,$$

$$\omega-\text{Im}\,\gamma(k^2,\lambda)+\text{Im}\,(a_1\sigma+a_2\tau+\cdots)=0.$$

We shall return in Chapter 4 to a discussion of the different solution types (i) through (iii) for bifurcating waves. For the record, we calculate the isotropy subgroups of solutions of type (i) and (ii). the isotropy subgroups are conjugate along orbits of the group action, so it suffices to calculate them at some fixed point along the orbit. Now $\sigma_\theta T_\gamma z_1 = e^{i(\theta+\gamma)z}1$ (take $k=1$ for this calculation) and $\sigma_\theta T_\gamma z_2 = e^{i(\theta-\gamma)}z_2$; so for an appropriate choice of θ and γ we may shift both z_1 and z_2 to the positive real axis. For solutions of type (i) we have $z_1 = z_2$, and the only symmetry is $P(Pz_1 = z_2, Pz_2 = z_1)$. For solutions of type (ii), say $z_1 \neq 0$ and $z_2 = 0$, the only symmetries are $\sigma_\theta T_{-\theta}$. This subgroup of transformations is isomorphic to $SO(1)$. Observe that for all other points in our 4-dimensional space $(z_1, z_2, \bar{z}_1, \bar{z}_2)$, the isotropy subgroup is trivial. Thus the solutions of types (i) and (ii) are characterized by having a larger isotropy subgroup than do general points in the space. We say such solutions are *isolated* in their strata (see Chapter 4).

A solution of type (ii) (say $z_2=0$) is of the form (to lowest order) $z_1 e^{i(kx+\omega t)}\varphi + \bar{z}_1 e^{-i(hx+\omega t)}\bar{\varphi} = 2 \operatorname{Re} z_1 e^{i(kx+\omega t)}\varphi$, and thus is a wave traveling to the left (if k and ω are of the same sign). On the other hand, a solution of type (i) $(z_1 = z_2)$ is of the form (taking $z_1 = z_2$ to be real and positive)

$$z_1 2 \operatorname{Re} (e^{i(kx+\omega t)} + e^{i(\omega t - kx)})\varphi,$$

and this has the waveform

$$\cos (kx + \omega t) + \cos (kx - \omega t) = 2 \cos kx \cos \omega t,$$

which is a standing wave.

We may picture these solutions as follows. The bifurcation equations are set in C^4, but since we want real solutions we restrict ourselves to the subspace of C^4 invariant under the operator J, where $J(z_1, z_2, z_3, z_4) = (\bar{z}_3, \bar{z}_4, \bar{z}_1, \bar{z}_2)$. A real mapping $F: C^4 \to C^4$ is equivariant with respect to J. The symmetry group of the problem acts on C^4 via the action σ_θ, T_γ, P. Each orbit under this action passes through the real subspace (x_1, x_2, x_1, x_2), that is, where $z_1 = x_1$ and $z_2 = x_2$ are real. The subgroup which leaves this real subspace invariant is generated by

$$T = \begin{pmatrix} -1 & 0 \\ 0 & -1 \end{pmatrix} \quad \text{and} \quad P = \begin{pmatrix} 0 & 1 \\ 1 & 0 \end{pmatrix}$$

(looking now only at the first two components). Therefore we may reduce the bifurcation problem to mappings $F: \mathbb{R}^2 \to \mathbb{R}^2$ which are equivariant with respect to T and P. As before, this leads to the general form

$$a\begin{pmatrix} x_1 \\ x_2 \end{pmatrix} + b\begin{pmatrix} x_1^3 \\ x_2^3 \end{pmatrix} = 0,$$

where a and b are functions of the invariants. The traveling wave solutions (type (i)), given by $r = s$, lie on the lines $x_1 = \pm x_2$. The standing waves ($r = 0$ or $s = 0$) lie on the x_1 or x_2 axes. When looked at in the reduced problem (the (x_1, x_2) plane), cf. Fig. 2.2) the rotating waves have a trivial isotropy subgroup; only when considered as a solution in the original space C^4 do they have a nontrivial isotropy subgroup.

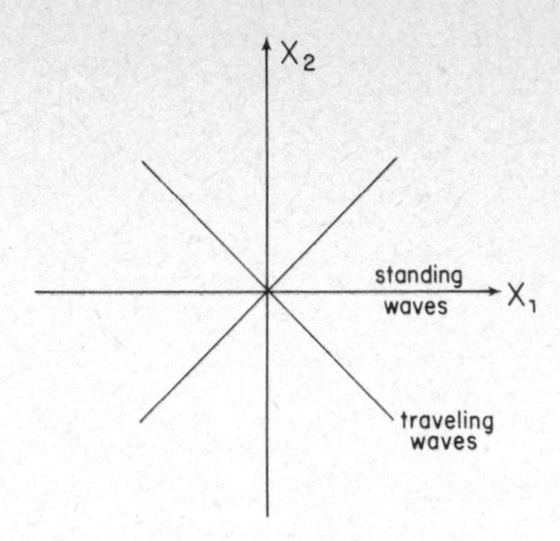

FIG. 2.2

A discussion of bifurcating waves on the surface of a sphere was given in [58]. Some partial results in case the kernel $\mathcal{N}_i$ transformed according to the representation $D^{(2)}$ were given, but the algebraic problem is in general much more complicated.

2.7. Stability of bifurcating waves. We discuss computational methods for determining the stability of bifurcating waves. Writing the bifurcation problem in the form

$$H(\lambda, w, u) = Ju + (\omega - 1)\frac{\partial u}{\partial s} - (L(\lambda) - L_0)u + R(\lambda, u) = 0,$$

we obtain the following eigenvalue problem for the Floquet exponents:

$$H_u w + \beta w = 0, \qquad w(0) = w(2\pi) = 0.$$

Some of the Floquet exponents may be zero due to the invariance of the equations under time translations or continuous space transformations, but if all the other Floquet exponents have negative real parts then the time periodic solutions are stable. If, as in the case of bifurcating waves on the line discussed in §2.6, the wave number k is allowed to vary, then the Floquet exponents obtained by the above procedure only test stability of the solutions relative to disturbances of the same wave number. One must therefore keep in mind that the Floquet analysis above does not give a complete resolution of the stability question in that case.

In [56, p. 82] it was shown that the stability of the bifurcating solutions in a neighborhood of the branch point is determined to lowest order by the Jacobian of the branching equations. Let $L(\varepsilon) = G_u(\lambda(\varepsilon), u(\varepsilon))$ be the linearized operator along a one-parameter branch $\lambda(\varepsilon)$, $u(\varepsilon)$, and let

$$P(\varepsilon) = \frac{1}{2\pi i}\int (z - L(\varepsilon))^{-1}\, dz,$$

where C encloses the origin and contains only the isolated zero eigenvalue of L_0. The stability of the bifurcating solutions is then determined (for small ε) by the eigenvalues of the finite dimensional operator $B(\varepsilon)=P(\varepsilon)L(\varepsilon)$. We proved ([56, Thm. 4.3]):

THEOREM 2.4. *Consider a general bifurcation problem* $G(\lambda, u)=0$. *Let the branching equations be denoted by* $F(\lambda, v)=0$. *Suppose these have a one-parameter family of solutions* $\lambda=\varepsilon^m\tau_0$, $v=\varepsilon^l\xi$, $\xi=\xi_0+\varepsilon\xi_1+\varepsilon^2\xi_2+\cdots$, *and suppose that*

$$F(\varepsilon^m\tau_0, \varepsilon^l\xi)=\varepsilon^kQ(\tau_0, \xi_0)+O(\varepsilon^{k+1})$$

where $k>\mathrm{Max}\,\{m, l\}$. *The reduced bifurcation equations are therefore* $Q(\tau, \xi)=0$. *Then the operator* $B(\varepsilon)=L(\varepsilon)P(\varepsilon)$ *is given by*

$$B(\varepsilon)=\varepsilon^{k-l}Q_\xi(\tau_0, \xi_0)+O(\varepsilon^{k-l+1}).$$

This theorem was proved with the time independent case in mind, but it holds equally well in the time periodic case. Accordingly, the sign of the critical eigenvalues can be determined from the eigenvalues of $Q_\xi(\tau_0, \xi_0)$.

Let us first rederive the stability results for the case of the Hopf bifurcation problem. Here the branching equations were

$$F_1=az=0, \qquad F_2=\bar{a}\bar{z}=0,$$

where $\sigma=|z|^2$ and $a=i(\omega-1)-\lambda\gamma'(0)+a_1\sigma+a_2\sigma^2+\cdots$. The Jacobian of these equations is

$$\frac{\partial(F_1, F_2)}{\partial(z, \bar{z})}=\begin{pmatrix} a+a_\sigma\sigma & a_\sigma z^2 \\ \bar{a}_\sigma\bar{z}^2 & \bar{a}+\bar{a}_\sigma\sigma \end{pmatrix}.$$

Since solutions are given by $a=0$, the Jacobian reduces to

$$\begin{pmatrix} a_\sigma\sigma & a_\sigma z^2 \\ \bar{a}_\sigma\bar{z}^2 & \bar{a}_\sigma\sigma \end{pmatrix}$$

along a solution. The eigenvalues of this matrix are 0 and $2\,\mathrm{Re}\,\sigma a_\sigma$, so the Floquet exponents are 0 and, to lowest order, $\beta=-2\,\mathrm{Re}\,\sigma a_\sigma$. Thus bifurcating solutions are stable if and only if $\mathrm{Re}\,a_\sigma=\mathrm{Re}\,a_1>0$. This condition is easily related to the direction of bifurcation. Taking real parts of the equation $a=0$, we get

$$-\lambda\,\mathrm{Re}\,\gamma'(0)+\mathrm{Re}\,a_1\sigma+\cdots=0, \qquad \lambda\approx\left(\frac{\mathrm{Re}\,a_1}{\mathrm{Re}\,\gamma'(0)}\right)\sigma.$$

(This is approximate since the coefficients a_j depend on λ.) Now $\mathrm{Re}\,\gamma'(0)>0$ by our assumption that the trivial solution becomes unstable as λ crosses 0. Therefore $\lambda>0$ if and only if $\mathrm{Re}\,a_1>0$, and supercritical solutions are stable, subcritical solutions, unstable.

In the case of bifurcating waves on the line, we have

$$F=1=(a+br)z_1, \quad F_2=(a+bs)z_2, \quad \bar{F}_1=(\bar{a}+\bar{b}r)\bar{z}_1, \quad \bar{F}_2=(\bar{a}+\bar{b}s)\bar{z}_2.$$

The Jacobian of the bifurcation equations at a solution of type (ii) ($z_2=0$,

$a+br=0$) is

$$\frac{\partial(F_1, \bar{F}_1, F_2, \bar{F}_2)}{\partial(z_1, \bar{z}_1, z_2, \bar{z}_2)} = \begin{pmatrix} B+b_\sigma r^2 & B+b_\sigma r^2 & 0 & 0 \\ \bar{B}+\bar{b}_\sigma r^2 & \bar{B}+\bar{b}_\sigma r^2 & 0 & 0 \\ 0 & 0 & a & 0 \\ 0 & 0 & 0 & \bar{a} \end{pmatrix}$$

where $B=(a_\sigma+b)r$. The eigenvalues of this matrix are 0, a, $\bar{a}$ and $2\,\mathrm{Re}\,\{(a_\sigma+b)r+b_\sigma r^2\}$. The situation is somewhat more complex than that above, and we defer a complete stability analysis to a future article.

CHAPTER 3

Equivariant Singularity Theory

3.1. Modules of equivariant mappings. In this chapter we present equivariant singularity theory and its application to symmetry breaking bifurcation problems. The application of singularity theory to bifurcation problems was developed in a series of articles by Golubitsky and Schaeffer [35], [36], [37]. In the examples, I have tied some of the phenomena which occur in the unfolding of singularities together with Michel's notion of critical orbits of group actions, discussed in Chapter 4. The critical orbits give rise to generic solutions of the bifurcation equations, and other, special solutions which appear as secondary bifurcations in the unfolding.

Let us recapitulate the notions of invariants and modules of equivariant mappings introduced in Chapter 2. Let $\mathcal{G}$ be a compact group acting linearly on $\mathbb{R}^n$. We denote points in $\mathbb{R}^n$ by x and the action of $\mathcal{G}$ by γx. A smooth mapping $F:\mathbb{R}^n \to \mathbb{R}^n$ is said to be equivariant if $\gamma F(x) = F(\gamma x)$. A function $h:\mathbb{R}^n \to \mathbb{R}$ is said to be invariant if $h(\gamma x) = h(x)$ for all $x \in \mathbb{R}^n$ and $\gamma \in \mathcal{G}$. The space of smooth invariant functions $h:\mathbb{R}^n \to \mathbb{R}$ forms a commutative ring with identity ($h = 1$) which we denote by I. Thus, if $h, g \in I$ then so do $h+g$ and hg. The space of smooth equivariant mappings $F:\mathbb{R}^n \to \mathbb{R}^n$ forms a module over I which we denote by $\mathscr{E}$. A Hilbert basis for a ring of invariant polynomials is a set of invariant polynomials $\sigma_1, \ldots, \sigma_n$ such that every invariant h is a polynomial in $\sigma_1, \ldots, \sigma_n$. For example, a Hilbert basis for the symmetric polynomials in $x_1 \cdots x_n$ is the set of elementary symmetric polynomials

$$\sigma_1 = x_1 + \cdots + x_n,$$

$$\sigma_2 = \sum_{1 \leq i < j \leq n} x_i x_j,$$

$$\vdots$$

$$\sigma_n = x_1 \cdots x_n.$$

A Hilbert basis for the ring r of polynomials on the real line which are invariant under the reflection $x \to -x$ is $\sigma(x) = x^2$: every even polynomial can be written in the form $h(x) = f(x^2)$.

As another example, consider the adjoint action of $GL(n)$ on the vector space of $n \times n$ matrices given by

$$\gamma \cdot A = PAP^{-1}, \qquad P \in GL(n).$$

A Hilbert basis for this action is the set of invariants

$$\operatorname{Tr} A, \quad II(a), \quad \ldots, \quad \det A.$$

These invariants are realized as the coefficients in the characteristic polynomials of A:

$$\det(\lambda I - A) = \lambda^n - (\mathrm{Tr}\, A)\lambda^{n-1} + II(A)\lambda^{n-2} + \cdots + (-1)^n (\det A) I.$$

Equivalently, one may take the invariants $\sigma_k(A) = \mathrm{Tr}\, A^k$, $k = 1, \ldots, n$, as a Hilbert basis.

As another example, let the group D_n act on the complex plane C by the operations

$$z \to \bar{z} \quad \text{and} \quad z \to e^{i\alpha} z \tag{3.1}$$

where $\alpha = 2\pi/n$. Then $z\bar{z}$ and $2\,\mathrm{Re}\, z^n = z^n + \bar{z}^n$ form a Hilbert basis for I. In fact, let $h(z, \bar{z}) \in I$ and

$$h(z, \bar{z}) = \sum a_{jk} z^j \bar{z}^k.$$

Then the invariance of h under the generators implies $h(z, \bar{z}) = h(z, \bar{z})$ and $h(e^{i\alpha} z, e^{-i\alpha}\bar{z}) = h(z, \bar{z})$. Since z and $\bar{z}$ must be regarded as independent variables, this implies $a_{jk} = a_{kj}$ and $a_{jk} = 0$ unless $j - k \equiv 0 \pmod n$. Thus a basis for I is given by

$$z^j \bar{z}^k + \bar{z}^j z^k = z^k \bar{z}^k (z^{nl} + \bar{z}^{nl})$$

where $j - k = nl \geqq 0$. We claim every one of these polynomials is a polynomial in $z\bar{z}$ and $z^n + \bar{z}^n$. This is clearly true for $l = 0, 1$. For $l > 1$ we have identity

$$z^{nl} + \bar{z}^{nl} = (z^n + \bar{z}^n)(z^{n(l-1)} + \bar{z}^{n(l-1)}) - z^n \bar{z}^n (z^{n(l-2)} + \bar{z}^{n(l-2)}).$$

The claim follows by induction on this recursion relation.

By a theorem of G. Schwarz [59], $u = z\bar{z}$ and $v = \mathrm{Re}\, z^n$ also generate the ring of invariant C^∞ functions.

A module $\mathscr{E}$ is said to be finitely generated if there exist vectors $x_1, \ldots, x_k$ in $\mathscr{E}$ such that every $x \in \mathscr{E}$ can be expanded as a linear combination

$$x = r_1 x_1 + \cdots + r_k x_k$$

with coefficients r_i in the ring. In this case the rank of $\mathscr{E}$ is said to be k. The module is said to be *free* if whenever $r_1 x_1 + \cdots + r_k x_k = 0$ we necessarily have $r_1 = \cdots = r_k = 0$. Since in general elements of a ring do not have multiplicative inverses, it is possible that a basis for $\mathscr{E}$ is not free, yet none of the generators can be replaced by a linear combination of the others.

Examples. 1. Let $\mathscr{G} = \{\mathrm{id}, \gamma\}$ act on the real line, where $\gamma x = -x$. In this case I consists of the even functions and $\mathscr{E}$ is generated by the function $h(x) = x$; it is a module of rank 1 over I generated by $\{x\}$.

2. When D_n acts on C as in Example 3.1, the generators of the module of equivariant functions are z and $\bar{z}^{n-1}$. Let $h(z, \bar{z})$ be given by

$$h = \sum_{j,k} a_{jk} z^j \bar{z}^k.$$

The equivariance conditions on h this time are $h(z, \bar{z}) = \overline{h(\bar{z}, z)}$ and $e^{i\alpha}h(z, \bar{z}) = h(e^{i\alpha}z, e^{-i\alpha}\bar{z})$. These lead to $a_{jk} = \bar{a}_{jk}$ and $a_{jk} = 0$ unless $j \equiv k+1 \pmod n$. The module of equivariant polynomials is generated by

$$(z\bar{z})^k z^{nl+1}, \quad l \geqq 0, \qquad (z\bar{z})^j \bar{z}^{nl-1}, \quad l \geqq 1.$$

Since $(z\bar{z})^k$ belongs to the ring I, the module is generated by $\{z^{nl+1}, l \geqq 0\}$ and $\{\bar{z}^{nl-1}, l \geqq 1\}$. We claim the generators of this module are z and $\bar{z}^{n-1}$. This follows by induction on the recursion relations

$$z^{ln+1} = (z^{ln} + \bar{z}^{ln})z - (z\bar{z})^{ln-1},$$
$$\bar{z}^{ln-1} = (z^{(l-1)n} + \bar{z}^{(l-1)n})\bar{z}^{n-1} - (z\bar{z})^{n-1}z^{(l-2)n+1}.$$

The general equivariant polynomial mapping is therefore

$$h(z, \bar{z}) = a(u, v)z + b(u, v)\bar{z}^{n-1}$$

where a and b are polynomials in u and v. This also holds for C^∞ equivariant maps by a result of Poenaru [4] which generalizes the result of Schwarz mentioned above.

Another module will figure in our singularity theory calculations. If $F(x)$ is an equivariant mapping, then $J(x, w) = F'(x)w$, where F' is the Jacobian, is equivariant also: $\gamma J(x, w) = J(\gamma x, \gamma w)$. In fact, since F is equivariant, $F(\gamma(x + tv)) = \gamma F(x + tv)$. Differentiating with respect to t and setting $t = 0$ we get $\gamma F'(x)v = F'(\gamma x)\gamma v$, hence the equivariance of J.

More generally, let $\mathcal{G}$ act on a manifold M via the action φ_g. Thus $\varphi_{g_1} \circ \varphi_{g_2} = \varphi_{g_1 g_2}$. There is a natural group action induced on $T^1(M)$ given by $\Phi_g(x, v) = (\varphi_g(x), \varphi_g'(x)v)$. (This action was already introduced in §2.1 in our discussion of the canonical action on vector fields.) Let F be an equivariant mapping of M, let $J(x, v) = F'(x)v$, where $v \in T_x(M)$, and define $F^1(x, v) = (F(x), J(x, v))$. Then $F^1: T^1 \to T^1$ and is equivariant with respect to Φ_g. In fact

$$\begin{aligned} \Phi_g F^1(x, v) &= \Phi_g(F(x), J(x, v)) \\ &= (\varphi_g \circ F, \varphi_g'(F(x))F'(x)v) \\ &= (F \circ \varphi_g, F'(\varphi_g(x))\varphi'(x)v) \\ &= F^1(\varphi_g(x), \varphi_g'(x)v) \\ &= (F^1 \circ \Phi_g)(x, v). \end{aligned}$$

(The identity $\varphi_g'(F(x))F'(x)v = F'(\varphi_g(x))\varphi_g'(x)v$ is established by differentiating the identity $\varphi_g \circ F = F \circ \varphi_g$ along a curve $x(t) \in M$, with $v = \dot{x}(t) \in T^1(M)$.) The mappings $J(x, v)$ form a module over the ring of invariant functions, denoted by $\mathcal{J}$. In the linear case, elements of $\mathcal{J}$ can be identified with matrices $T(x)$ such that $\gamma T(x) = T(\gamma x)\gamma$. Returning to our example D_n, we look for polynomials $h(z, \bar{z}, w, \bar{w})$ which are linear in w, $\bar{w}$ and equivariant with respect to the action of D_n on $C \times C$:

$$h(z, \bar{z}, w, \bar{w}) = h(\bar{z}, z, \bar{w}, w),$$
$$e^{i\alpha}h(z, \bar{z}, w, \bar{w}) = h(e^{i\alpha}z, e^{-i\alpha}\bar{z}, e^{i\alpha}w, e^{-i\alpha}\bar{w}).$$

Expanding

$$h=\sum a_{jk}z^j\bar{z}^k w + b_{jk}z^j\bar{z}^k\bar{w},$$

we find

$$a_{jk}=\bar{a}_{jk} \quad \text{and} \quad a_{jk}=0 \quad \text{unless } j=k \ (\text{mod } n),$$
$$b_{jk}=\bar{b}_{jk} \quad \text{and} \quad a_{jk}=0 \quad \text{unless } j=k+2 \ (\text{mod } n).$$

The generators of $\mathcal{J}$ in this case are

$$\{w, z^n w, z^2\bar{w}, \bar{z}^{n-2}\bar{w}\}. \tag{3.2}$$

Again these are also the generators of the ring of C^∞ equivariant mappings.

3.2. Unfoldings. A *bifurcation* problem with symmetry group $\mathcal{G}$ is a one-parameter family $F_\lambda(x)=F(x, \lambda)$ of elements of $\mathcal{E}$ [35]. We say two bifurcation problems F_1 and F_2 are *$\mathcal{G}$-contact equivalent* if there exists a triple $\{T(x, \lambda), X(x, \lambda), \Lambda(\lambda)\}$ such that $F_1(x, \lambda)=T(x, \lambda)F_2(X(x, \lambda), \Lambda(\lambda))$ where

(i) $X\in\mathcal{E}$, that is, $\gamma X(x, \lambda)=X(\gamma x, \lambda)$;

(ii) T is a nonsingular $n\times n$ matrix in $\mathcal{J}$, that is, $\gamma T(x, \lambda)=T(\gamma x, \lambda)\gamma$;

(iii) $\det\|\partial X^i/\partial x^j\|_{x=\lambda=0}>0$ and $\Lambda'(0)>0$.

$\mathcal{G}$-contact equivalence is an equivalence relation, and if $F_2\in\mathcal{E}$ then $TF_2\circ(X, \Lambda)\in\mathcal{E}$.

The ordinary definitions of unfolding theory carry over directly to the case where symmetry is present. An l-parameter $\mathcal{G}$-unfolding of F is a multi-parameter family $G(x, \lambda, \alpha)\in\mathcal{E}$ such that $G(x, \lambda, 0)=F(x, \lambda)$ and $\alpha\in\mathbb{R}^l$. A second k-parameter unfolding $H(x, \lambda, \beta)$, $\beta\in\mathbb{R}^k$, is said to factor through G if there is a mapping $\beta\to\alpha(\beta)$ such that $H(x, \lambda, \beta)$ and $G(x, \lambda, \alpha)$ are $\mathcal{G}$-contact equivalent with the contact equivalence depending smoothly on β:

$$H(x, \lambda, \beta)=T(x, \lambda, \beta)G(X(x, \lambda, \beta), \Lambda(\lambda, \beta), \alpha(\beta)).$$

G is called a universal unfolding of F if every $\mathcal{G}$-unfolding of F factors through G, that is, if every $\mathcal{G}$-unfolding of F is contact equivalent to some G.

If G is a $\mathcal{G}$-universal unfolding of F, then every $\mathcal{G}$-equivariant perturbation of F is $\mathcal{G}$-equivalent to some $G(x, \lambda, \alpha(\beta))$ with α depending smoothly on the perturbation parameters β. Thus, the structure of the zero set of perturbations of F is qualitatively the same as that of G and undergoes the same structural changes as that of $\mathcal{G}$. We investigate the qualitative structure of the solution set of $F=0$ by embedding the reduced bifurcation equations $Q=0$ in a universal unfolding of Q. Then F is $\mathcal{G}$-equivalent to that universal unfolding.

Let F be an element of $\mathcal{E}$ and consider the orbit $\mathcal{O}_F$ through F under $\mathcal{G}$-contact equivalence:

$$\mathcal{O}_F=\{F_1\in\mathcal{E} \mid F_1 \text{ is } \mathcal{G}\text{-contact equivalent to } F\}.$$

The "tangent space" to $\mathcal{O}_F$ at F in the module $\mathcal{E}$ is formally obtained by differentiating along arbitrary one-parameter curves through F. Thus, let us suppose we have

$$F_1(x, \lambda, t) = T(x, \lambda, t)F(X(x, \lambda, t), \Lambda(\lambda, t)),$$

where $T(x, \lambda, 0) = I$, $X(x, \lambda, 0) = x$ and $\Lambda(\lambda, 0) = \lambda$. Let δ be the operation $d/dt\,|_{t=0}$; then

$$\delta F_1 = (\delta T)F(x, \lambda) + F_x \cdot \delta X + F_\lambda\, \delta\Lambda,$$

where F_x is the Jacobian of F with respect to x and $F_\lambda = \partial F/\partial\lambda$.

Let $\tilde{T}_F$ denote the submodule spanned by $(\delta T)F$ and $F_x \cdot \delta X$ as δT and δX range over the generators of $\mathcal{J}$ and $\mathcal{E}$:

$$\tilde{T}_F = \langle(\delta T)F, F_x \cdot \delta X\rangle = \langle T_i F, \ldots, T_k \cdot F, F_x X_1, \ldots, F_x \cdot X_l\rangle$$

where $\langle T_1, \ldots, T_k\rangle = \mathcal{J}$ and $\langle X_1, \ldots, X_l\rangle = \mathcal{E}$. The $\mathcal{G}$-equivariant mapping $F(x, \lambda)$ is said to have finite codimension if the quotient module $\mathcal{E} \mid \tilde{T}_F$ has finite dimension (over the ring I). The actual codimension of F is computed a little differently. We define

$$T_F = \tilde{T}_F + \mathcal{E}_\lambda \left\{\frac{\partial F}{\partial \lambda}\right\}$$

where $\mathcal{E}_\lambda$ is the ring of functions of λ alone. T_F is not really a module, since we do not allow multiplication of $\partial F/\partial\lambda$ by arbitrary elements of the ring I. The $\mathcal{G}$-codimension of F is defined to be the dimension of $\mathcal{E}/T_F$ where T_F is considered as a vector subspace of $\mathcal{E}$.

We can picture the situation as shown in Fig. 3.1. T_F is the tangent space to the orbit $\mathcal{O}_F$ and $\mathcal{E}/T_F$ is the complementary subspace or, better, the orbit space. The subspace $\mathcal{E}/T_F$ parametrizes the nonequivalent classes of mappings F, that is, the unfolding of F. Each F' near F is equivalent to some mapping in the unfolding of F; in other words, the orbit $\mathcal{O}_{F'}$ intersects $\mathcal{E}/T_F$ at some point.

THEOREM 3.1 [36]. *Let $F \in \mathcal{E}$, let F have finite $\mathcal{G}$-codimension l, and let the generators of $\mathcal{E}/T_F$ be $F_1(x, \lambda), \ldots, F_l(x, \lambda)$. Then a universal unfolding of f is*

$$G(x, \lambda, \alpha) = F(x, \lambda) + \alpha_1 F_1(x, \lambda) + \cdots + \alpha_l F_l(x, \lambda).$$

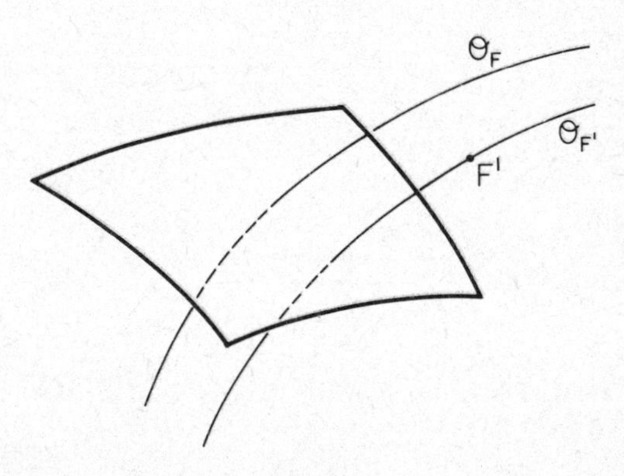

FIG. 3.1

3.3. Universal unfolding of the Z_2 singularity. Consider one-dimensional bifurcation problems equivariant with respect to the action $\gamma x=-x$. As we saw above, the general equivariant mapping is of the form

$$F(x, \lambda)=a(u, \lambda)x$$

where $u=x^2$. The module $\mathcal{J}$ consists of scalars $m(x)$ such that $\gamma m(x)= m(\gamma x)\gamma$, thus $-m(x)=m(-x)(-1)$; therefore $\mathcal{J}$ consists of even functions, and $\mathcal{J}$ is generated by 1. The generators of $\tilde{T}_F$ are therefore $1\cdot F$ and $F_x\cdot x$:

$$\tilde{T}_F=\langle a(u, \lambda)x, (a_u\cdot 2x+a)x\rangle=\langle ax, ua_ux\rangle,$$

$$T_F=\{(r_1(u, \lambda)a+r_2(u, \lambda)ua_u+\sigma(\lambda)a_\lambda(u, \lambda))x\}.$$

For example, if $F(x, \lambda)=x^3-\lambda x=(u-\lambda)x$, then $\tilde{T}_F=\langle(u-\lambda)x, ux\rangle= \langle\lambda x, ux\rangle$. An arbitrary element H of $\mathscr{E}$ can be written in the form

$$H=a(u, \lambda)x=(a_0+\lambda a_1(u, \lambda)+ua_2(u, \lambda))x\equiv a_0x \pmod{\tilde{T}_F}$$

where a_0 is a fixed constant. Since $F_\lambda=-x$ and a_0 is constant, $H\equiv O \pmod{T_F}$, and therefore the codimension of F is zero. This differs from the ordinary codimension of F, which is 2 [35]. The universal unfolding of $x^3-\lambda x$ is $x^3-\lambda x+px^2+q$. (See [35], also [57].)

Now consider $F=x^3-\lambda^2x=(u-\lambda^2)x$. This time $\tilde{T}_F=\langle ux, \lambda^2x\rangle$, and an arbitrary element of $\mathscr{E}$ can be expanded as

$$\begin{aligned}H(x, \lambda)&=H(u, \lambda)x\\&=(h(0,0)+(h(0, \lambda)-h(0,0))+(h(u, \lambda)-h(0, \lambda)))x\\&=h_0x+h_1\lambda x+\lambda^2h_2(\lambda)x+h_3(u, \lambda)ux\\&\equiv h_0x+h_1\lambda x \pmod{\tilde{T}_F}.\end{aligned}$$

Since $F_\lambda=-2\lambda x$, $H\equiv h_0x \pmod{T_F}$, the codimension of F is 1 and the universal unfolding of F is $x^3-\lambda^2x+\alpha x$.

3.4. Universal unfolding of the D_n singularity. We now consider a more ambitious example: We take D_n acting on C via the two-dimensional representation (3.1). We have seen that I is generated by $u=z\bar{z}$ and $v=z^n+\bar{z}^n$, and that $\mathscr{E}$ is generated by z and $\bar{z}^{n-1}$, so a general element of $\mathscr{E}$ is of the form $F(z, \bar{z}, \lambda)=a(u, v, \lambda)z+b(u, v, \lambda)\bar{z}^{n-1}$.

LEMMA 3.2. *The generators of the submodule* $\tilde{T}_F$ *are*

$$\begin{aligned}F_1&=az+b\bar{z}^{n-1},\\F_2&=(av+bu^{n-1})z-au\bar{z}^{n-1},\\F_3&=(au+bv)z-bu\bar{z}^{n-1},\\F_4&=u^{n-2}b\cdot z+a\bar{z}^{n-1},\\F_5&=(a+2ua_u+nva_v)z+(2ub_u+(n-1)b+nvb_v)\bar{z}^{n-1},\\F_6&=(va_u+2nu^{n-1}a_v+(n-1)u^{n-2}b)z\\&\quad+(a+2nu^{n-1}b_v+vb_u)\bar{z}^{n-1}.\end{aligned}$$

(See [37] for the case $n=3$.)

Proof. From (3.2) the generators of $\tilde{T}_F$ corresponding to the generators of $\mathcal{J}$ are F, z^nF, $z^2\bar{F}$ and $\bar{z}^{n-2}\bar{F}$; these give rise to the first four generators above. For example, F_2 is obtained by

$$\begin{aligned} F_2 = z^nF = (v-\bar{z}^n)F &= vF - \bar{z}^n(az + b\bar{z}^{n-1}) \\ &= vF - au\bar{z}^{n-1} - b\bar{z}^{n-1}(v-z^n) \\ &= vF - (au+vb)\bar{z}^{n-1} + u^{n-1}bz \\ &= (av + bu^{n-1})z - au\bar{z}^{n-1}. \end{aligned}$$

The others are obtained similarly. The Jacobian of F is $\{F_z, F_{\bar{z}}\}$. In fact, let $z(t)$, $\bar{z}(t)$ be a curve in C and let $w=\dot{z}$, $\bar{w}=\dot{\bar{z}}$; then $\delta F = F_z w + F_{\bar{z}}\bar{w}$. F_5 and F_6 are obtained as $F_5 = F_z z + F_{\bar{z}} z$, $F_6 = F_z \bar{z}^{n-1} + F_{\bar{z}} z^{n-1}$.

THEOREM 3.3. *Let* $F = (Au + \alpha\lambda)z + (Cu + Dv + \beta\lambda)\bar{z}^{n-1}$ *and assume*

$$AD \neq 0, \quad \alpha \neq 0, \quad \alpha C \neq \frac{n}{n-2} A\beta.$$

Then codim $F = 2$ *and*

$$\mathcal{E}/T_F = \langle uz, \bar{z}^{n-1} \rangle.$$

(This is a slightly specialized version of a result due to Golubitsky and Schaeffer [37, Thm. 3.4].)

Proof. We begin with the slightly more general case $G = (Au + Bv + \alpha\lambda)z + (Cu + Dv + \beta\lambda)\bar{z}^{n-1}$ and then, for simplicity, restrict ourselves to the case $B = 0$. The generators of $\tilde{T}_F$ in this case are

$$\begin{aligned} &F_1 = (Au + Bv + \alpha\lambda)z + (Cu + Dv + \beta\lambda)\bar{z}^{n-1}, \\ &F_2 = (v(Au + Bv + \alpha\lambda) + u^{n-1}(Cu + Dv + \beta\lambda))z - u(Au + Bv + \alpha\lambda)\bar{z}^{n-1}, \\ &F_3 = (u(Au + Bv + \alpha\lambda) + v(Cu + Dv + \beta\lambda))z - u(Cu + Dv + \beta\lambda)\bar{z}^{n-1}, \\ &F_4 = u^{n-2}(Cu + Dv + \beta\lambda)z + (Au + Bv + \alpha\lambda)\bar{z}^{n-1}, \\ &F_5 = (3Au + (n+1)Bv + \alpha\lambda)z + ((n+1)Cu + (2n-1)Dv + \beta\lambda(n-1))\bar{z}^{n-1}, \\ &F_6 = (Av + u^{n-2}(2nBu + (n-1)Cu + (n-1)(D + \beta\lambda)))z \\ &\qquad + (Au + (B+C)v + \alpha\lambda + 2nu^{n-1}D)\bar{z}^{n-1}. \end{aligned}$$

Now we claim that $\langle u\bar{z}^{n-1}, v\bar{z}^{n-1}, \lambda\bar{z}^{n-1} \rangle \equiv \langle uz, \lambda z \rangle \bmod \tilde{T}_F$ if $B = 0$ and the conditions in the theorem are satisfied. In fact, $F_1 \equiv F_4 \equiv F_5 \equiv 0 \pmod{\tilde{T}_F}$ and this gives

$$\begin{aligned} (Cu + Dv + \beta\lambda)\bar{z}^{n-1} &\equiv -(Au + \alpha\lambda)z, \\ (Au + Bv + \alpha\lambda)\bar{z}^{n-1} &\equiv -u^{n-2}(Cu + Dv + \beta\lambda)z, \\ ((n+1)Cu + (2n-1)Dv + (n-1)\beta\lambda)\bar{z}^{n-1} &\equiv (3Au + (n+1)Bv + \alpha\lambda)z. \end{aligned}$$

These equations may be solved for $u\bar{z}^{n-1}$, $v\bar{z}^{n-1}$, $\lambda\bar{z}^{n-1}$ in terms of the module $\langle uz, \lambda z \rangle$ over I provided the matrix (recall that $B = 0$)

$$\begin{pmatrix} A & 0 & \alpha \\ C & D & \beta \\ (n-1)C & (2n-1)D & (n-1)\beta \end{pmatrix}$$

is invertible. This matrix is now equivalent to

$$\begin{pmatrix} A & 0 & \alpha \\ 0 & AD & A\beta-\alpha C \\ 0 & 0 & \frac{n}{2-n}A\beta+\alpha C \end{pmatrix},$$

which is nonsingular when the conditions in the theorem are satisfied.

Therefore a general H in $\mathscr{E}$ can be written

$$\begin{aligned} H = az + b\bar{z}^{n-1} &\equiv h(u, v, \lambda) + h_0\bar{z}^{n-1} \\ &= (h_0(\lambda) + uh_1 + vh_2)z + h_0\bar{z}^{n-1} \\ &\equiv (uh_1 + vh_2)z + h_0\bar{z}^{n-1} \quad (\text{mod } T_F) \end{aligned}$$

since $h_0(\lambda)z \in T_F$ if $\alpha \neq 0$ (for then $(1/\alpha)\,\partial F/\partial\lambda = z \in T_F$).

We now replace vz by uz mod T_F. From $F_6 \equiv 0$ we get

$$\begin{aligned} Avz &\equiv -(Au + (B+C)v + \alpha\lambda + 2nDu^{n-1})\bar{z}^{n-1} \quad (\text{mod } \langle uz, \tilde{T}_F\rangle) \\ &\equiv q\lambda z \quad (\text{mod } \langle uz, \tilde{T}_F\rangle) \end{aligned}$$

where q is a constant. Therefore a general term in $\langle vz\rangle$ is of the form

$$\begin{aligned} \chi(u, v, \lambda)vz \equiv \chi(u, v, \lambda)\lambda z &\equiv (\chi_0(\lambda) + u\chi_1 + v\chi_2)\lambda z \\ &\equiv \chi_2\lambda vz \quad (\text{mod } \langle uz, T_F\rangle) \end{aligned}$$

since $\chi_0(\lambda)\lambda z \in T_F$. But from $F_2 \equiv 0$ we get (recall that $B = 0$)

$$\alpha\lambda vz \equiv u(Au + \alpha\lambda)\bar{z}^{n-1} \equiv u(\mu u + \nu\lambda)z \equiv 0 \quad (\text{mod } \langle uz, \tilde{T}_F\rangle).$$

Since we have reduced H (mod T_F) to

$$H \equiv \sigma_1 uz + \sigma_2\bar{z}^{n-1},$$

codim $F = 2$ and $\mathscr{E}/T_F = \langle uz, \bar{z}^{n-1}\rangle$.

THEOREM 3.4. *The mapping* $Q = \lambda z + b\bar{z}^2$ *is structurally stable; that is,* codim $Q = 0$.

This theorem was proved in [37] by a method different from ours. We may interpret the result in the following way. If the reduced bifurcation equations are given by $Q = 0$, then the solution structure of the full bifurcation equations is the same as that of $Q = 0$. In other words, the reduced bifurcation equations give the full qualitative picture.

Proof. We must show that a general $F = a(\lambda, u, v)z + b(\lambda, u, v)\bar{z}^2$ is zero modulo T_Q. From Lemma 3.2 the generators of $\tilde{T}_Q$ are:

$$\begin{aligned} Q_1 &= \lambda z + b\bar{z}^2, & Q_4 &= ubz + \lambda\bar{z}^2, \\ Q_2 &= (\lambda v + bu^2)z - \lambda u\bar{z}^2, & Q_5 &= \lambda z + 2b\bar{z}^2, \\ Q_3 &= (\lambda u + bv)z - bu\bar{z}^2, & Q_6 &= 2ubz + \lambda\bar{z}^2. \end{aligned}$$

From $b^{-1}(Q_5 - Q_1) = \bar{z}^2$ we see that $\bar{z}^2$, hence any term of the form $b(\lambda, u, v)\bar{z}^2$

is zero modulo $\tilde{T}_Q$. Therefore $F \equiv a(\lambda, u, v)z$ modulo $\tilde{T}_Q$. Furthermore, $b^{-1}Q_4 = uz + \lambda\bar{z}^2 \equiv uz \bmod \tilde{T}_Q$ and $b^{-1}Q_3 = ((\lambda/b)u + v)z - u\bar{z}^2 \equiv vz \bmod \tilde{T}_Q$. Therefore $\langle uz, vz\rangle$ belongs to $\tilde{T}_Q$. Now we may write

$$F \equiv (a_0(\lambda) + ua_1(\lambda, u, v) + va_2(\lambda, u, v))z,$$

but $a_1(\lambda, u, v)uz + a_2(\lambda, u, v)vz \in \langle uz, vz\rangle$, and therefore $F \equiv a_0(\lambda)z \bmod \tilde{T}_Q$. Finally, $\partial Q/\partial\lambda = z$, so $F \equiv a_0(\lambda)z \equiv 0 \bmod T_Q$.

3.5. Bifurcation analysis for D_3. As we showed in §2.4 (bifurcation in the presence of the rotation group), the bifurcation problem for the $D^{(2)}$ representation of $O(3)$ reduces to an analysis of the branching problem with D_3 symmetry. Let us begin by assuming the reduced bifurcation equations are $Q = \lambda z + b\bar{z}^2$. By Theorem 3.4, Q is structurally stable, so in this case the full bifurcation picture is qualitatively given by the zero set $Q = 0$.

Putting $z = re^{i\theta}$ we have

$$Q(z, \lambda) = re^{i\theta}(\lambda + bre^{-3i\theta}).$$

So $Q = 0$ on $z = 0$ and on $\lambda + bre^{-3i\theta} = 0$. Since b is real we must have

$$\lambda = -br, \qquad \theta = \frac{2\pi k}{3},$$

or

$$\lambda = br, \qquad \theta = \frac{(2k+1)\pi}{3}.$$

The two solution types bifurcate on opposite sides of criticality. We depict the situation for, say, $b > 0$ in Fig. 3.2.

The Jacobian of Q is

$$\begin{pmatrix} Q_z & Q_{\bar{z}} \\ \bar{Q}_z & \bar{Q}_{\bar{z}} \end{pmatrix} = \begin{pmatrix} \lambda & 2b\bar{z} \\ 2bz & \lambda \end{pmatrix}.$$

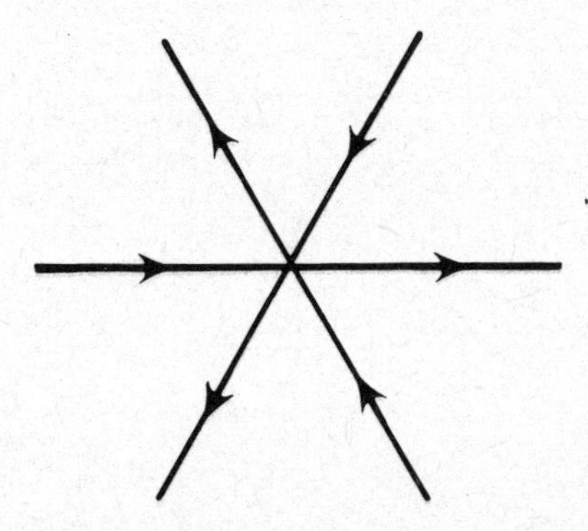

FIG. 3.2. *Bifurcation for $Q = 0$, $b > 0$. The arrows indicate the motion of the solutions for increasing λ. The arrows are reversed if* sgn b *is reversed.*

The determinant of this matrix is $\lambda^2 - 4b^2r^2 = -3b^2r^2 < 0$, so the branching solutions are always unstable.

Now consider $F = (\lambda - u)z + (u + Dv)\bar{z}^2$. This time codim $F = 2$ and the universal unfolding is $F = (\lambda - Au)z + (u + Dv + E)\bar{z}^2$ where A and E are the unfolding parameters. When $E = 0$, the only solution is the critical orbits which are those lying along the rays on which $z/\bar{z}^2$ is real. (The theory of critical orbits will be developed in Chapter 4.) The critical orbits are given by $\arg z = k\pi/3$ for integral k. In fact, if we try to solve $a = b = 0$ we get ($z = re^{i\theta}$):

$$a = \lambda - Au = 0, \qquad b = u + Dv = 0,$$

hence

$$\lambda = Ar^2, \qquad r^2(1 + Dr\cos 3\theta) = 0.$$

The second equation has no solution for small r.

Now let us locate the critical orbits. These are generated (under the group action) by

(i) $z = r,\ \bar{z}^2 = r^2,\ \bar{z} = rz,\ v = r^3,$
(ii) $z = -r,\ \bar{z}^2 = r^2,\ \bar{z} = -r,\ v = -r^3.$

In case (i) $F = ar + br^2$ so the branching equations are

$$a + rb = (\lambda - Ar^2) + r^3 + Dr^4 + Er = 0,$$
$$\lambda = -Er + Ar^2 - r^3 - r^4.$$

In case (ii) we get $a - rb = 0$,

$$\lambda = Er + Ar^2 + r^3 - r^4.$$

When $E = 0$ the two critical orbits bifurcate on the same side of criticality, supercritically if $A > 0$ and subcritically if $A < 0$. The symmetry group of these critical solutions is in each case Z_2. One can work out their stability properties, and one finds that near the branch point their eigenvalues are of opposite sign.

As E changes from zero, a secondary bifurcation takes place, the new solutions having smaller (in fact, trivial) isotropy subgroups. Let us solve for the secondary solutions when $E \neq 0$. The equations were given above, and are

$$\lambda = Ar^2, \qquad r^2 + Dr^3 \cos 3\theta = E.$$

Turning to the second equation, we plot (Fig. 3.3) $r^2 + Dr^3$ and $r^2 - Dr^3$ (assuming $D > 0$). We find that $r^2 - Dr^3$ has a local maximum of $4/27D^2$ at $r = 2/3D$. Therefore $0 < E < 4/27D^2$ is a necessary condition that the equation $r^2 + r^3D\cos 3\theta = E$ have a solution. In that case the equation describes a smooth curve in the z-plane, and r varies between r_1 and r_2. Having solved the second equation, we put $\lambda = Ar^2$. Since r never vanishes, λ never changes sign. This curve of special solutions intersects the generic solutions (i.e., those along the critical orbits) more or less as we have indicated in Fig. 3.4. The stability analysis of the secondary solutions was carried out in [37]; the stability of the secondary solutions depends on the sign of D. Their stability assignment

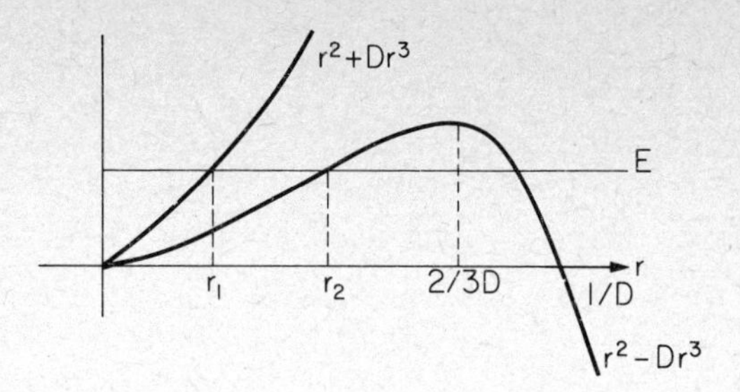

FIG. 3.3

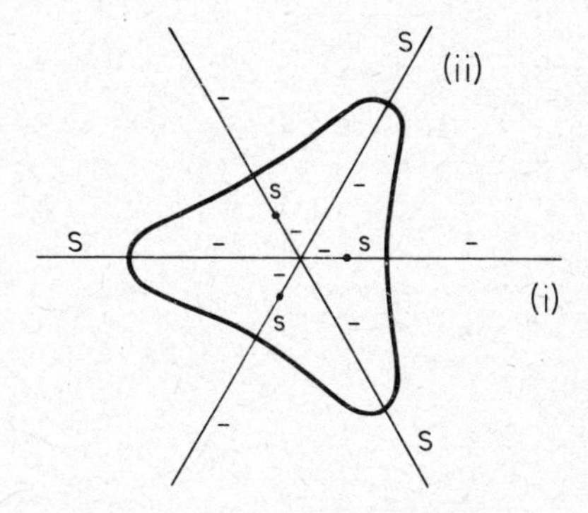

FIG. 3.4. *Projection of the bifurcation diagram into the z-plane. Distance from the origin represents distance along the bifurcating branch. Stability designations are: s for stable, – for eigenvalue of opposite sign, u for two unstable modes*

alternates between s and – in going around the circuit; it changes at every point of intersection with the generic solutions.

CHAPTER 4

Critical Orbits of Linear Group Actions

4.1. Orbits and strata. The equivariant bifurcation equations take the form $\lambda x = F(x, \lambda)$ where F is equivariant with respect to the action $(x, \gamma) \to \gamma x$ of some symmetry group $\mathcal{G}$ acting on the kernel $\mathcal{N}$ of the linearized equations. For applications to problems in classical mechanics, the group $\mathcal{G}$ is a group of rigid motions acting by orthogonal transformations—that is, transformations which preserve the Euclidean inner product. Thus, $\langle \gamma x, \gamma y \rangle = \langle x, y \rangle$. But even in other cases, $\mathcal{G}$ often acts as a group of unitary transformations which leave a scalar product invariant.

L. Michel has observed [43], [44] that in such situations it is possible to characterize a certain class of generic solutions of equivariant bifurcation equations entirely in terms of the group action. To see how this situation arises, let us consider some simple examples.

Example 4.1. Let $\mathcal{G}$ be the group D_2 acting in the plane which is generated by the reflections $\gamma_1(x, y) = (-x, y)$ and $\gamma_2(x, y) = (x, -y)$, and let $\mathcal{E}$ be the family of equivariant maps of $\mathbb{R}^2$ into itself. then $f = (f_1, f_2) \in \mathcal{E}$ implies

$$-f_1(x, y) = f_1(-x, y) = f_1(-x, -y),$$
$$-f_2(x, y) = f_2(x, -y) = f_2(-x, -y).$$

Now consider mappings of $\mathcal{E}$ restricted to the unit circle $x^2 + y^2 = 1$. It is easily seen that the points $(0, \pm 1)$ and $(\pm 1, 0)$ are critical points for $\mathcal{E}$ in the sense that every $f \in \mathcal{E}$ points normal to the unit circle at each of these points. (See Fig. 4.1.) In particular, if I denotes the set of invariant real valued functions in the plane, then $F \in I$ implies $\nabla F \in \mathcal{E}$. Since ∇F is normal to the unit circle at these points, $(\nabla F)(p) = \lambda p$, so these points are critical points for every invariant function F restricted to the unit circle.

Example 4.2. Consider the group S^1 acting on the unit sphere S^2 in $\mathbb{R}^3$ by rotations about the z-axis. The set of fixed points of this action consists precisely of the z-axis. If f is equivariant, then $\gamma f(\mathbf{x}) = f(\gamma \mathbf{x})$, and so $f(\mathbf{x})$ is fixed under the group action whenever $\mathbf{x}$ is. Consequently $f(\mathbf{x})$ lies on the z-axis whenever $\mathbf{x}$ does, and any equivariant mapping must be normal to the sphere at the fixed points of the action—that is, at the north and south poles. These two points are consequently critical points of the action on the unit sphere.

DEFINITION 4.3. Let V be an inner product space with inner product $\langle \cdot, \cdot \rangle$. Let $\mathcal{G}$ be a group of isometries acting linearly on V and let M be a $\mathcal{G}$-invariant submanifold of V. Denote by $\mathcal{E}$ the set of all equivariant mappings of M into V. The critical points of the action of $\mathcal{G}$ on M are those $\mathbf{x} \in M$ at which every $f \in \mathcal{E}$ is normal to M.

The ρ sphere in V, $S_\rho = \{x \mid \langle x, x \rangle = \rho\}$, is a $\mathcal{G}$-invariant submanifold of V.

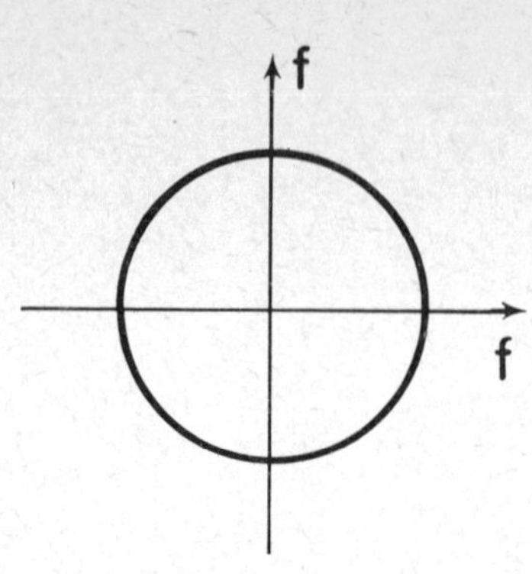

FIG. 4.1

The tangent space $T_X(S_\rho)$ is the set

$$T_x(S_\rho) = \{y \mid \langle x, y \rangle = 0\},$$

and the normal space at x is the set $N_x = \{\lambda x \mid \lambda \in \mathbb{R}\}$. Therefore, in the case $M = S_\rho$, x is a critical point of the action if for every $f \in \mathscr{E}$ we have $f(x) = \lambda x$ for some λ.

Since the gradient of an invariant function is an equivariant map, the critical points of the action will be critical points on M of *every* invariant function on V restricted to M.

The subgroup $G^x = \{\gamma \mid \gamma x = x\}$ is called the *isotropy* (or sometimes *little* or *stabilizer*) subgroup of x. If x and y lie on the same orbit their isotropy subgroups are conjugate, for if $y = hx$ then $G^y = hG^x h^{-1}$. A *stratum* is a subset of v all of whose points have conjugate isotropy subgroups.

In Example 4.1 there are three strata: the points $(\pm 1, 0)$ with isotropy subgroup $\{\text{id}, \gamma_2\}$, the points $(0, \pm 1)$ with isotropy subgroup $\{\text{id}, \gamma_1\}$ and all others, whose isotropy subgroup is trivial. In Example 4.2 there are two strata: the north and south poles, whose isotropy subgroup is the full group S^1, and the remaining points, whose isotropy subgroup is trivial.

In both examples the critical points are precisely those which are *isolated in their strata*: that is, all neighboring points except those on the same orbit have a strictly smaller isotropy subgroup. (We partially order subgroups of $\mathscr{G}$ by saying that $H_1 < H_2$ if H_1 is conjugate to a subgroup of H_2.)

We introduce some useful terminology. Let $\pi : V \times V \to V$ be defined by $\pi(x, y) = x$, and let $\mathscr{G}$ act on $V \times V$ by $\gamma(x, y) = (\gamma x, \gamma y)$. Then π is an equivariant map: $\gamma\pi = \pi\gamma$. A mapping $f : V \to V$ is regarded as a section of $V \times V$. Let $G_f =$ graph of $f = \{(x, f(x)) \mid x \in V\}$. Then G_f is an invariant submanifold of $V \times V$ if and only if f is equivariant. (By an invariant submanifold we mean a submanifold S such that $\gamma S \subset S$ for $\gamma \in \mathscr{G}$.) Let M be an invariant submanifold of V, and let $N(M)$ be its normal bundle. Then $N(M)$ is an invariant submanifold of $V \times V$. In fact, $(x, y) \in N(M)$ if and only if $\langle w, y \rangle = 0$ for all $w \in T_x(M)$. But then $\langle \gamma w, \gamma y \rangle = \langle w, y \rangle = 0$ and $\gamma w \in T_{\gamma x}(M)$, so $\gamma y \in N_{\gamma x}(M)$.

A critical point of $f(x)$ restricted to M is a point $x \in M$ such that $G_f \cap \pi^{-1}(x) \subset N_M \cap \pi^{-1}(x)$. Since the group action leaves G_f and N_M invariant,

$$\gamma(G_f \cap \pi^{-1}(x)) \subset \gamma(N_M \cap \pi^{-1}(x));$$

thus

$$G_f \cap \pi^{-1}(\gamma x) \subset N_M \cap \pi^{-1}(\gamma x)$$

so γx is critical whenever x is, and the critical points occur in orbits.

4.2. Theorems of L. Michel. L. Michel has proved the following result [44]:

THEOREM 4.4. *Let M be an invariant submanifold, and for each $x \in M$ assume $N_G(G^x)/G^x$ is discrete, where $N_G(G^x)$ is the normalizer subgroup of G^x in G. Then the critical orbits are precisely those which are isolated components of their strata.*

The normalizer subgroup $N_G(H)$ is defined to be $N_G(H) = \{a \in G \mid aHa^{-1} = H\}$. The condition that $N_G(G^x)/G^x$ be discrete was not explicitly contained in [44]. It is not needed for gradient vector fields but it is necessary for equivariant vector fields. I had originally introduced the condition in connection with Theorem 4.7. During the conference N. Dancer observed that such a condition was necessary in this case, too, and supplied me with the counterexample below. Michel has since written me pointing out that he was aware that additional conditions are needed in the case of equivariant vector fields. One such condition, for example, is that the isotropy subgroup G_x contain a Cartan subgroup of G. (See Rev. Mod. Phys. 52 (1980), p. 642.)

We first prove

LEMMA 4.5. *Let M be an invariant submanifold. For every $x \in M$ there is a neighborhood U, $x \in U \subset M$, such that $G^z < G^x$ for every $z \in U$.*

The inequality here refers to inclusion up to conjugation.

Proof. Since M is an invariant submanifold of V, it inherits a metric ρ from V which is invariant under the group action. Let $\mathcal{O}_x$ be the orbit through x (see Fig. 4.2). Given $z \in U \backslash \mathcal{O}_x$, there is a unique $y \in \mathcal{O}_x$ such that $\rho(z, y)$ is a minimum. Define the retraction map r by $r(z) = y$. Then r is equivariant. In

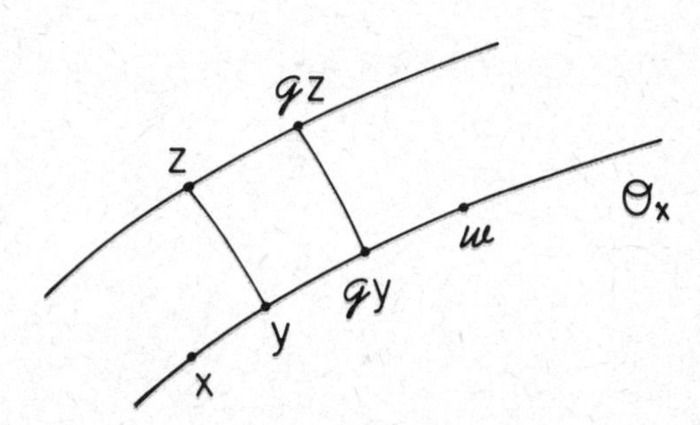

FIG. 4.2

fact, suppose $r(gz)=w$. Then by definition of r and the invariance of ρ,

$$\rho(z, g^{-1}w)=\rho(gz, w)\leqq\rho(gz, gy)=\rho(z, y).$$

Since y gives the unique minimum of $\rho(z,\cdot)$ on $\mathcal{O}_x$, we conclude that $y=g^{-1}w$, hence $w=gy$.

Now if $r(z)=x$ then $G^z\subset G^x$, for if $g\in G^z$ we have $gx=gr(z)=r(gz)=r(z)=x$ so $g\in G^x$. For other nearby points $x''\notin\mathcal{O}_x$ we have $r(x'')=gx$ for some g since $r(x'')$ lies on the orbit $\mathcal{O}_x$, so $G^{x''}\subset G^{gx}=gG^xg^{-1}$. Finally, for $y\in\mathcal{O}_x$, $y=gx$ and $G^y=gG^xg^{-1}$. Thus, in all cases and for all $y\in U$, G^y is conjugate to a subgroup of G^x.

Proof of Theorem 4.4. Let U be an open set of M and let $U\times V$ be the trivial bundle over U. Let $(x, F(x))$ be an invariant section of $U\times V$ and consider $h_t(x)=\pi(x+tF(x))$ for small t, where π is the projection of $U\times V$ onto U. Since π is equivariant,

$$gh_t(x)=g\pi(x+tF(x))=\pi(gx+tgF(x))=\pi(gx+tF(gx))=h_t(gx),$$

so h_t is equivariant. Therefore the isotropy subgroup G^x leaves $h_t(x)$ fixed as well (see Fig. 4.3). If $F(x)$ is an invariant section of $U\times V$ which does not lie in the normal bundle of M, then $h_t(x)$ is a family of orbits lying near the orbit $\mathcal{O}_x$ with the same isotropy subgroup ($G^{h_t(x)}\subset G^x$ by the lemma). Hence if $\mathcal{O}_x$ is not a critical orbit, it is embedded in a stratum with the same isotropy subgroup. Equivalently, if the orbit $\mathcal{O}_x$ is isolated in its stratum, then it is critical.

We must still show that $h_t(x)$ does not lie in $\mathcal{O}_x$, and here is where we use the normalizer condition. We have already shown that $G^{h_t(x)}=G^x$. If $h_t(x)\in\mathcal{O}_x$ then $h_t(x)=p(t)x$ for some $p(t)\in G$, and $G^{h_t(x)}=p(t)G^xp^{-1}(t)=G^x$, so $p(t)\in N_G(G^x)$. But if $h_t(x)\neq x$ (i.e., if $F(x)\neq\lambda x$), then $p(t)\notin G^x$.

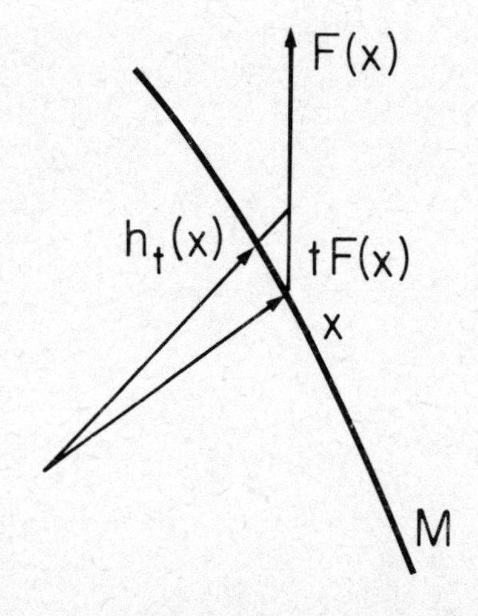

FIG. 4.3

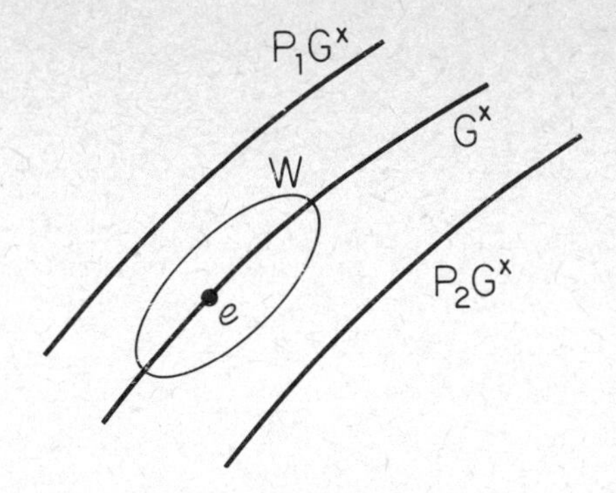

FIG. 4.4

Since $N_G(G^x)/G^x$ is discrete, we can write $N_G(G^x) = \bigcup_i p_i G^x$ for some $p_i \in N_G(G^x)$. The picture is given in Fig. 4.4. There is a neighborhood W of e in g such that $W \cap N_G(G^x) \subset G^x$. Thus for $p(t)$ near e (t near zero), $p(t)$ cannot belong to $N_G(G^x)$ unless it belongs to G^x. Therefore $h_t(x) \notin \mathcal{O}_x$, and we have shown that if $\mathcal{O}_x$ is an isolated component of its stratum, then $\mathcal{O}_x$ is critical.

Conversely, suppose the orbit through x is not isolated in its stratum. Let S be the stratum containing x. We give a proof that $\mathcal{O}_x$ is not a critical orbit under the assumption that S is locally (in a neighborhood of the orbit $\mathcal{O}_x$) a smooth submanifold of M and that the complement of S is itself a stratum. Let $r^{-1}(x) = \{y \mid r(y) = x\}$ and let U be a neighborhood of x. Then $r^{-1}(x) \cap U$ is, for sufficiently small U, a smooth submanifold of M which intersects the orbit $\mathcal{O}_x$ orthogonally and whose codimension is equal to the dimension of the orbit $\mathcal{O}_x$ (see Fig. 4.5).

We construct a smooth mapping F with the following properties:

(1) F maps $r^{-1}(x) \cap U$ into itself and F leaves S and S^c invariant.

(2) F is identically the identity outside some subdomain $r^{-1}(x) \cap U' \subset r^{-1}(x) \cap U$.

(3) $F(x) \neq \lambda x$ for any scalar λ.

Such a mapping may be constructed by constructing a vector field v in $r^{-1}(x) \cap U$ such that v is tangent to S, $v(x) \neq 0$ and v vanishes identically

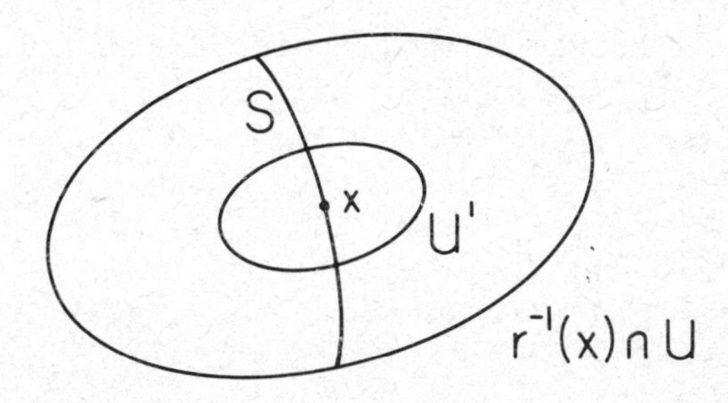

FIG. 4.5

outside U'. Let φ_t be the flow generated by v. Then for small ε the mapping $F(y) = \varphi_\varepsilon(y)$ satisfies the conditions (1)–(3) above.

Now extend F to be an equivariant mapping of M into itself by defining $F(gy) = gF(y)$ for all points $y \in r^{-1}(x) \cap U$. Every point in the tubular neighborhood of the orbit $\mathcal{O}_x$ can be represented in this way. We now have constructed an equivariant map F for which $F(x) \neq \lambda x$ for any scalar λ. Thus the orbit $\mathcal{O}_x$ is not a critical orbit.

This concludes the proof of Theorem 4.4.

Here is the counterexample which shows that the normalizer condition is needed. Let S^3 be the unit sphere in $\mathbb{R}^4$ and let $S^1 \times S^1$ act on S^3 by separate rotations in x^1, x^2 and x^3, x^4. Thus an element $T(\alpha, \beta) \in S^1 \times S^1$ is of the form

$$T(\alpha, \beta) = \begin{pmatrix} R(\alpha) & 0 \\ 0 & R(\beta) \end{pmatrix}$$

where $R(\alpha)$ is a 2×2 rotation matrix. Let F be a rotation in the (x^1, x^2) plane which leaves x^3 and x^4 fixed:

$$F = \begin{pmatrix} R_0 & 0 \\ 0 & I \end{pmatrix}$$

where R_0 is a 2×2 rotation and I is the 2×2 identity matrix. F is clearly equivariant. The mapping $h_t(x) = \pi(\mathbf{x} + tF(\mathbf{x}))$ is a rotation about the equator $u = v = 0$, which is an isolated component of its stratum. Its isotropy subgroup is S^1 while the normalizer subgroup is the full group $S^1 \times S^1$ itself, since the latter is abelian. The quotient group is S^1. The equator is not a critical orbit of F since F does not point normal to the sphere along the equator.

I would like to thank M. Golubitsky and L. Michel for their valuable comments on the material in this section.

4.3. A trace criterion for critical orbits. Let $\mathcal{G}$ act on V and for a given $x \in V$ let $V^x = \{y \mid G^y = G^x\}$. If $\mathcal{G}$ acts linearly, then V^x is a linear subspace of V. If $\dim V^x = 1$, then x is isolated in its stratum. For from the proof of Lemma 4.5, the isotropy subgroup of every point $y \in r^{-1}(x)$ is a subgroup of G^x. Yet, if x is not isolated in its stratum, then for all nearby $y \in r^{-1}(x)$, we must have G^y conjugate to G^x; hence $G^y = G^x$. If $\dim V^x = 1$, this is impossible.

This leads to one possible method for determining critical orbits. Let T_g denote a linear representation of G on a vector space V and let $F(x)$ be equivariant with respect to T_g. Given $x \in V$, the representation T_g restricted to the isotropy subgroup G^x decomposes into a direct sum of irreducible representations of G^x, and T_g contains the identity representation precisely $\dim V^x$ times. We have proved:

THEOREM 4.6. *Let T_g be a linear representation of a Lie group $\mathcal{G}$ on a vector space V; let $x \in V$, and let G^x be the isotropy subgroup of x. If $T_g|_{G^x}$ contains the identity representation precisely once, then x is a critical point of the action.*

Here is a second, very quick proof. If F is equivariant then $G^x \subseteq G^{F(x)}$. If

$\dim V^x = 1$ then $G^x = G^{F(x)}$ and $F(x) = \lambda x$ for some scalar λ; therefore x is a critical point of the action.

A similar result, but with an additional assumption on the bifurcation mapping F, was obtained by G. Cicogna [22].

Let H be a subgroup of G. the normalizer subgroup of H is the subgroup $N_G(H) = \{g \mid g \in G \cdot gHg^{-1} = H\}$. Clearly H is a normal subgroup of $N_G(H)$. We may prove a partial converse to Theorem 4.6.

THEOREM 4.7. *Let $N_G(G^x)/G^x$ be discrete. Then a necessary condition that x be a critical point is that $T_g|_{G^x}$ contain the identity representation precisely once.*

Proof. We claim that if $\dim V^x > 1$ then there is a neighborhood U of x such that $V^x \cap \mathcal{O}_x \cap U = \{x\}$. If $y \in \mathcal{O}_x$ and $y = hx$ then $G^y = hG^xh^{-1}$, so $V^x \cap \mathcal{O}_x = \{y \mid y = hx$ for some $h \in N_G(G^x)\}$. Let the cosets of G^x in $N_G(G^x)$ be $\{h_1G^x, h_2G^x, \dots, h_kG^x\}$. Then $N_G(G^x) = \bigcup_i h_iG^x$, and there is an open set W of e in G such that $W \cap N_G(G^x) = G^x$.

Now for a sufficiently small neighborhood U of x, $y \in \mathcal{O}_x \cap U$ and $y \neq x$ if and only if $y = hx$ for some $h \in W \backslash G^x$. But then $h \notin N_G(G^x)$, so $y \notin V^x$. This proves our first claim.

Since $\dim V^x > 1$ and V^x does not intersect $\mathcal{O}_x$, V^x must intersect $r^{-1}(x)$ in a smooth submanifold containing x. But then x is not isolated in its stratum, so x is not a critical point by Theorem 4.4.

The number of times the identity representation is contained in $T_g|_{G^x}$ is easily computed in terms of the characters of G. Let $\chi(g) = \operatorname{Tr} T_g$ be the character of T_g and let $\chi^{(\nu)}(g)$ be ths characters of the irreducible representations of G^x. The number of times the jth representation is contained in $T_g|_{G^x}$ is ([46, p. 77])

$$\mathcal{A}_j = \frac{1}{|G^x|} \sum_{g \in G^x} \chi(g)\overline{\chi^{(j)}(g)}.$$

(If G^x is a continuous group, the sum must be replaced by an invariant integral and $|G^x|$ by

$$|G^x| = \int_{G^x} d\mu(g)$$

where $\mu(g)$ is the invariant measure on G^x. I am assuming G^x is compact.) In particular, since the character of the identity representation is identically one, the identity representation is contained precisely $\mathcal{A}_0$ times:

$$\mathcal{A}_0 = \frac{1}{|G^x|} \sum_{g \in G^x} \chi(g) = \frac{1}{|G^x|} \sum_i g_i\chi_i,$$

where the second sum is over the conjugacy classes in G^x, g_i is the number of elements in the ith conjugacy class, and χ_i is the value of χ on the ith conjugacy class.

4.4. Critical orbits for representations of $SO(3)$. Let us carry out this program (at least in part) for the rotation group. Denote the irreducible representations of $O(3)$ by $D^{(l+)}$ and $D^{(l-)}$. The representation $D^{(l-)}$ carries the reflection in $O(3)$ ($\mathbf{x}\rightarrow -\mathbf{x}$) into the matrix $(-I)$ where I is the $(2l+1)\times(2l+1)$ identity matrix. Let us consider, for simplicity, only the representations $D^{(l+)}$ and, further, let us consider only the group $SO(3)$ of pure rotations.

A complete list of the point groups of the first kind is given by the cyclic groups C_m, the dihedral groups D_m, $m\geqq 2$, and the tetrahedral, octahedral, and icosahedral group T, O, and Y. (See Miller [46, pp. 27–32].) Each point symmetry group H of $SO(3)$ consists of a discrete set of rotations about a finite set of axes $v_1, v_2, \ldots$. A rotation of magnitude $2\pi/n$ about an axis v is denoted by $C_{2\pi/n}(v)$, and v is said to be an n-fold axis. If H is a subgroup of $SO(3)$ with axes $v_1, v_2, \ldots, v_k$, then conjugation of H by g takes these axes into $gv_1, \ldots, gv_k$, and gHg^{-1} is the group of rotations about the axes $gv_1, \ldots, gv_k$. Consequently the normalizer subgroup of H is the subgroup of $SO(3)$ which permutes all the n-fold axes among themselves. In the case of the subgroups D_m, T, O and Y, the normalizer subgroup is the group itself, but in the case C_m the normalizer subgroup is S^1. Therefore $N_{SO(3)}(H)/H$ is discrete for $H=D_m$, T, O, or Y. There is one subgroup remaining, and that is S^1, the subgroup of rotations about a fixed axis. Its normalizer is the group of all rotations in $SO(3)$ which leave that axis fixed, hence the original group S^1 plus rotations of π about any axis perpendicular to the invariant axis. In this case, it is clear that the cosets of S^1 in $N_{SO(3)}(S^1)$ are S^1 and gS^1 where g is a rotation of π about an orthogonal axis; hence the quotient group is discrete.

Furthermore, T, O, Y and S^1 are maximal proper subgroups of $SO(3)$—that is, they are contained in no larger subgroup. therefore, if $D^l(g)$ is a representation of $SO(3)$ and x is fixed under one of these groups, then that group is in fact the isotropy subgroup of x. We may therefore apply our method to any of the subgroups T, O, Y or S^1. For example, let us consider the case O:

THEOREM 4.8. *The values of l for which D^l restricted to O contains the identity representation precisely once are*

$$l=0, 4, 6, 8, 9, 10, 13, 14, 15, 17, 19, 23.$$

I will sketch the proof. It relies on computations for the rotation group and its point subgroups which can be found in Hammermesh [38, pp. 337–339].

The conjugacy classes of $O(3)$ are labelled by the magnitude φ of the rotation, and the character of the lth representation is

$$\chi^{(l)}(\varphi)=\frac{\sin(l+1/2)\varphi}{\sin\varphi/2}. \tag{4.1}$$

The octahedral subgroup O consists of pure rotations only and contains 24 elements in 5 conjugacy classes (see Hammermesh [38, p. 339]):

$$e,\quad C_3(8),\quad C_4^2(3),\quad C_2(6),\quad C_4(6).$$

Here e denotes the identity and C_n denotes a rotation of $2\pi/n$ about an axis; the numbers in parentheses denote the number of elements in the associated conjugacy class. The elements C_2 and C_4^2 are both rotations through an angle

π; they are conjugate in the full orthogonal group but not by elements of O, so they lie in distinct conjugacy classes in O. The decomposition of $D^{(l)}$ into sums of irreducible representations of O is given in a table by Hammermesh [38, p. 339]. In his notation, A_1 is the identity representation, A_2 is an alternating one-dimensional representation, E is a two-dimensional representation, and F_1 and F_2 are three-dimensional representations. The decompositions for $l=0$ through 5 are:

$$\begin{bmatrix} l=0 & A_1, \\ l=1 & F_1, \\ l=2 & E+F_2, \\ l=3 & A_2+F_1+F_2, \\ l=4 & A_1+E+F_1+F_2, \\ l=5 & E+2F_1+F_2. \end{bmatrix}$$

Thus $D^{(l)}$ restricted to O contains A_1 precisely once for $l=0$, 4, 6. To determine the splitting for other values of l we need some additional facts from group representation theory. The representation of a group G on itself given by $g \to L_g$ where $L_g h = gh$ is called the *left regular representation* of G. If G is a finite group with elements $g_1, \ldots, g_n$, each L_{g_i} will act as a permutation on the elements, and $\chi(g) = \operatorname{Tr} L_g =$ the number of elements fixed by L_g. Clearly $\chi(e)=|G|$ and if $g \neq e$ (e is the identity of G). The left regular representation can be decomposed into a direct sum of irreducible representation of G with multiplicity equal to the dimension of the corresponding irreducible representation. In the case of O, the regular representation is therefore

$$A_1+A_2+2E+3F_1+3F_2.$$

Now consider $D^{(12)}$. Since 12 is the least common multiple of 2, 3 and 4, $\chi_i^{(l)}=1$ for all the conjugacy classes except the identity, so

$$\chi^{(12)}(g)=\begin{cases}25 & \text{if } g=e, \\ 1 & \text{if } g \neq e.\end{cases}$$

Therefore on O we have $\chi^{(12)}=1+\chi_{\mathrm{reg}}$ where χ_{reg} is the character of the regular representation, and so $D^{(12)}$ restricted to O decomposes into

$$D^{(12)}|_O = A_1+\mathrm{reg} = 2A_1+A_2+2E+3F_1+3F_2.$$

Similarly,

$$D^{(12m)} = A_1+m(\mathrm{reg}), \quad \text{and for } k<12, \quad D^{(12m+k)} = m(\mathrm{reg})+D^{(k)}. \tag{4.2}$$

Finally, it is a simple matter to check from (4.1) that

$$\chi^{(l)}\left(\frac{2\pi}{n}\right)+\chi^{(11-l)}\left(\frac{2\pi}{n}\right)=0$$

and

$$\chi^{(l)}(e)+\chi^{(11-l)}(e)=24;$$

so that

$$D^{(l)}+D^{(11-l)}=(\text{reg})=A_1+A_2+2E+3F_1+3F_2.$$

Now we are ready to roll. From (4.3), $D^{(11-l)}$ contains the identity representation once if and only if $D^{(l)}$ does not. This gives $l=8$, 9, 10 in Theorem 4.8. From (4.2), we see that the remaining possible values are of the form $l=12+k$ where $D^{(k)}$ does not contain A_1.

For the values of l listed above, every bifurcation problem for which the null space transforms like $D^{(l)}$ has a bifurcating solution with symmetry group O.

COROLLARY 4.9. *Let $G(\lambda, u)=0$ be any regular nonlinear funcational equation with a bifurcation point at $u=0$, $\lambda=0$. Let $\mathcal{N}=\text{Ker } G_u(0,0)$. If $\mathcal{N}$ transforms under the rotation group $O(3)$ according to the representation D^{l+}, and l is one of the integers given in Theorem 4.8 (excluding $l=0$), then there exists a branching solution $(\lambda, u(\lambda))$ near $(0,0)$ with octahedral symmetry. Furthermore, these are the only values of l for which every bifurcation problem has branching solutions with octahedral symmetry.*

$l=0$ is excluded, since for $l=0$ one obtains only rotationally symmetric solutions. The kernel $\mathcal{N}$ consists of a one-dimensional subspace of rotationally invariant functions.

Warning. The values of l are given in Theorem 4.8 are those values of l for which bifurcating solutions with octahedral symmetry *must* exist. That is, those values of l are sufficient to guarantee the existence of solutions with octahedral symmetry. The solutions thus obtained are the generic or universal solutions for the problem. To find out when solutions with octahedral symmetry are *possible*, one must calculate those values of l for which $D^{(l)}|_D$ contains the identity representation at least once. These are given by:

THEOREM 4.10. *A necessary condition for the existence of bifurcating solutions with octahedral symmetry is that $\mathcal{N}$ transform according to D^l for one of the following values of l:*

$$l=4, 6, 8, 9, 10, \text{ or } \geqq 12.$$

Example 4.11. In §3.5 we considered the case where the dihedral group D_3 acted on $\mathbb{R}^2$ via its two-dimensional irreducible representation. We saw that the canonical form of the equivariant bifurcation equations were

$$G(z, \bar{z}, \lambda)=a(u, v, \lambda)z+b(u, v, \lambda)\bar{z}^2.$$

The invariants are $u=z\bar{z}$ and $v=\text{Re } z^3$. Writing $z=re^{i\theta}$ we can put g in the form $are^{i\theta}+br^{n-1}e^{-i(n-1)\theta}$. The critical points of this action are the rays for which $\sin n\theta=0$, for the symmetry group of the vectors $e^{i\pi/n}$ is Z_2 while the symmetry group of all other vectors is the trivial group. But if $\sin n\theta=0$, then $n\theta=k\pi$ for some integer k and $\bar{z}^{n-1}=\pm r^{n-2}z$, so on the critical rays

$$G=(a\pm r^{n-2}b)z.$$

Thus g points in the normal direction along the critical rays. If the bifurcation problem satisfies the Hopf transversality condition (the critical eigenvalues cross the imaginary axis with nonzero speed as λ crosses zero), then $(\partial a/\partial\lambda)(\lambda, 0, 0)|_{\lambda=0} = a_0 > 0$. Thus a takes the form $a = a_0\lambda + a_1 u + a_2 v + \cdots$. Moreover, on the critical points, $u = r^2$ and $v = \pm r$ so the bifurcation equations reduce to

$$C(\lambda, r) = a_0\lambda + c_1 r^2 + c_2 r^n + \cdots = 0.$$

This is readily solved for, say, λ as a function of r.

We saw in Chapter 2 that bifurcation in the presence of the $D^{(2)}$ representation of $O(3)$ reduces to this same bifurcation problem for $n = 3$. In that case the eigenvalues of the corresponding diagonal matrix are given by

$$d_1 = z + \bar{z} = 2r\cos\theta,$$

$$d_2 = \mu z + \mu^2\bar{z} = 2r\cos\left(\theta + \frac{2\pi}{3}\right),$$

$$d_3 = \mu^2 z + \mu\bar{z} = 2r\cos\left(\theta - \frac{2\pi}{3}\right).$$

The critical points of the S_3 action lie along the rays $\theta = k\pi/3$, k an integer. In each case two of the eigenvalues are $-r$ and the third is $2r$. Thus the corresponding matrix A is some permutation of

$$A = \begin{pmatrix} -r & 0 & 0 \\ 0 & -5 & 0 \\ 0 & 0 & 2r \end{pmatrix}.$$

The isotropy subgroup of such an a is $O(2) \times Z_2$. The Z_2 action comes from the invariance of the above A, for example, under the action by

$$\begin{pmatrix} 1 & 0 & 0 \\ 0 & 1 & 0 \\ 0 & 0 & -1 \end{pmatrix}.$$

Therefore the critical orbits of the $D^{(2)}$ action are the axisymmetric solutions.

References

[1] A. AMBROSETTI AND P. RABINOWITZ, *Dual variational methods in critical point theory and applications*, J. Funct. Anal., 14 (1973), pp. 349–381.

[2] A. BAHRI, *Une méthode perturbative en théorie de Morse et groupes d'homotopie des ensembles de niveaux*, Thèse de doctorat d'Etat, Univ. Pierre et Marie Curie, Paris, 1981.

[3] A. BAHRI AND H. BERESTYCKI, *Points critiques de perturbations de fonctionelles paires et applications*, C.R. Acad. Sci. Paris, Ser. A., 291 (1980), pp. 189–192.

[4] ———, *Existence d'une infinité de solutions périodiques pour certains systèmes Hamiltoniens en presence d'un terme de forcing*, C.R. Acad. Sci. Paris, Ser. A., 292 (1981), pp. 315–318.

[5] ———, *A perturbation method in critical point theory and applications*, Trans. Amer. Math. Soc. 267 (1981), pp. 1–32.

[6] ———, *Forced vibrations of super quadratic Hamiltonian systems*, Technical report, Laboratoire d'Analyse Numérique, 4 Place Jussieu, Paris.

[7] A. BAHRI AND H. BRÉZIS, *Periodic solutions of a nonlinear wave equation*, Proc. Royal Soc. Edinburgh, 85A (1980), pp. 313–320.

[8] V. BENCI, *Some critical point theorems and applications*, Comm. Pure Appl. Math., 33 (1980), pp. 146–171.

[9] ———, *A geometrical index theory for the group S^1 and some applications to research of periodic solutions of ordinary differential equations*, to appear.

[10] V. BENCI AND P. H. RABINOWITZ, *Critical point theorems for indefinite functionals*, Invent. Math., 53 (1979), pp. 241–273.

[11] H. BERESTYCKI, *Le nombre de solutions de certains problèmes semilinéaires elliptiques*, J. Funct. Anal., 40 (1981), pp. 1–29.

[12] H. BERESTYCKI AND P. L. LIONS, *Existence of stationary states in nonlinear scalar field equations*, in Bifurcation Phenomena in Mathematical Physics and Related Topics, C. Bardos and D. Bessis, eds., D. Reidel, Dordrecht, Holland, 1980, pp. 264–292.

[13] J. P. BOUJOT, J. P. MORERA AND R. TEMAM, *An optimal control problem related to the equilibrium of a plasma in a cavity*, App. Math. Optim., 2 (1975), pp. 97–129.

[14] H. BREZIS, J. M. CORON, AND L. NIRENBERG, *Free vibrations of a nonlinear wave equation and a theorem of P. Rabinowitz*, Comm. Pure Appl. Math., 33 (1980), pp. 667–684.

[15] H. BREZIS AND L. NIRENBERG, *Forced vibrations for a nonlinear wave equation*, Comm. Pure Appl. Math., 31 (1978), pp. 1–30.

[16] F. E. BROWDER, *Nonlinear eigenvalue problems and group invariance*, in Functional Analysis and Related Fields, F. E. Browder, ed., Springer-Verlag, New York, 1970, pp. 1–58.

[17] ———, *Infinite dimensional manifolds and nonlinear elliptic eigenvalue problems*, Ann. Math., 82 (1965), pp. 459–477.

[18] F. BUSSE, *The stability of finite amplitude cellular convection and its relation to an extremum principle*, J. Fluid Mech., 30 (1967), pp. 625–650.

[19] ———, *Patterns of convection in spherical shells*, J. Fluid Mech., 72 (1975), pp. 67–85.

[20] E. BUZANO AND M. GOLUBITSKY, *Bifurcation involving the hexagonal lattice*, Tech. Report 56, Arizona State Univ., Tempe, AZ, 1981.

[21] P. CHOSSAT, *Bifurcation and stability of convective flows in a rotating or not rotating spherical shell*, SIAM J. Appl. Math., 37 (1979), pp. 624–647.

[22] G. CICOGNA, *Symmetry breakdown from bifurcation*, Lett. Nuovo Cimento, 31 (1981), pp. 600–602.

[23] F. H. CLARKE AND I. EKELAND, *Hamiltonian trajectories having prescribed minimal period*, Comm. Pure Appl. Math., 33 (1980), pp. 103–116.

[24] C. V. COFFMAN, *A minimum maximum principle for a class of nonlinear integral equations*, J. d'Analyse Math., 22 (1969), pp. 391–419.

[25] E. N. DANCER, *On the existence of bifurcating solutions in the presence of symmetries*, Proc. Royal Soc. Edinburgh, 85A (1980), pp. 321–336.

[26] ———, *An implicit function theorem with symmetries and its application to nonlinear eigenvalue problems*, Bull. Austral. Math. Soc., 21 (1980), pp. 81–91.

[27] I. EKELAND, *Periodic solutions of Hamiltonian equations and a theorem of P. Rabinowitz*, J. Differential Equations, 34 (1979), pp. 523–534.

[28] I. EKELAND AND R. TEMAM, *Convex Analysis and Variational Problems*, North-Holland, Amsterdam, 1976.

[29] G. B. ERMENTROUT AND J. D. COWAN, *Large scale spatially organized activity in neural nets*, SIAM J. Appl. Math., 38 (1980), pp. 1–21.

[30] ———, *A mathematical theory of visual hallucination patterns*, Biol. Cybernet., 34 (1979), pp. 137–150.

[31] E. R. FADELL AND P. H. RABINOWITZ, *Generalized cohomological index theories for Lie group actions with an application to bifurcation questions for Hamiltonian systems*, Invent. Math., 45 (1978), pp. 139–174.

[32] L. E. FRAENKEL AND M. S. BERGER, *A global theory of steady vortex rings in an ideal fluid*, Acta Math., 132 (1974), pp. 13–51.

[33] M. GELL-MANN AND Y. NE'EMAN, *The Eightfold Way*, Benjamin, New York, 1964.

[34] W. M. GIBSON AND B. R. POLLARD, *Symmetry Principles in Elementary Particle Physics*, Cambridge Univ. Press, Cambridge, 1976.

[35] M. GOLUBITSKY AND D. SCHAEFFER, *A theory for imperfect bifurcation via singularity theory*, Comm. Pure Appl. Math., 32 (1979), pp. 21–98.

[36] ———, *Imperfect bifurcation in the presence of symmetry*, Comm. Math. Phys., 67 (1979), pp. 205–232.

[37] ———, *Bifurcation with O(3) symmetry including applications to the Bénard problem*, Comm. Pure Appl. Math., 35 (1982), pp. 81–111.

[38] M. HAMERMESH, *Group Theory and Its Applications to Physical Problems*, Addison-Wesley, Reading, MA, 1962.

[39] J. A. HEMPEL, *Multiple solutions for a class of nonlinear boundary value problems*, Indiana Univ. Math. J., 20 (1971), pp. 983–996.

[40] M. HERSCHKOWITZ-KAUFMAN AND T. ERNEUX, *The bifurcation diagram of model chemical reactions*, Ann. New York Acad. Sci., 316 (1979), pp. pp. 296–313.

[41] K. HOFFMAN AND R. KUNZE, *Linear Algebra*, Prentice-Hall, Englewood Cliffs, NJ, 1971.

[42] G. KNIGHTLY AND D. SATHER, *Applications of group representations to the buckling of spherical shells*, in Applications of Nonlinear Analysis in the Physical Sciences, H. Amman et al., eds., Pitman, London, 1981.

[43] L. MICHEL, *Nonlinear group action. Smooth action of compact Lie groups on manifolds*, in Statistical Mechanics and Field Theory, R. N. Sen and C. Weil, eds., Israel Univ. Press, Jerusalem, 1972, pp. 133–150.

[44] ———, *Simple mathematical models of symmetry breaking, application to particle physics*, in Mathematical Physics and Physical Mathematics, Polish Scientific Publishers, Warsaw, 1976, pp. 251–262.

[45] L. MICHEL AND L. A. RADICATI, *The geometry of the octet*, Ann. Inst. H. Poincaré Sect. A, 18 (1973), pp. 185–214.

[46] W. MILLER, *Symmetry Groups and Their Applications*, Academic Press, New York, 1972.

[47] W. M. NI, *Some minimax principles and their applications in nonlinear elliptic equations*, J. d'Analyse Math., 37 (1980), pp. 248–275.

[48] L. NIRENBERG, *Variational and topological methods in nonlinear problems*, Bull. Amer. Math. Soc., 4 (1981), pp. 267–302.

[49] R. S. PALAIS, *Ljusternik–Schnirelmann theory on Banach manifolds*, Topology, 5 (1966), pp. 115–132.

[50] ———, *Critical point theory and the minimax principle*, Proc. Symp. Pure Math., 15 American Mathematics Society, Providence, RI, 1970, pp. 185–212.

[51] V. POENARU, *Singularités C^∞ en présence de symétrie*, Lecture Notes in Mathematics 510, Springer-Verlag, Heidelberg, 1976.

[52] P. H. RABINOWITZ, *Variational methods for nonlinear eigenvalue problems*, in Eigenvalues of Nonlinear Problems, Ediz. Cremonese, Rome, 1974.

[53] ———, *Free vibrations of a semi-linear wave equation*, Comm. Pure Appl. Math., 31 (1978), pp. 31–68.

[54] ———, *Periodic solutions of Hamiltonian systems*, Comm. Pure Appl. Math., 31 (1978), pp. 157–184.

[55] D. H. SATTINGER, *On global solutions of nonlinear hyperbolic equations*, Arch. Rational Mech. Anal., 30 (1968), pp. 148–172.

[56] ———, *Group Theoretic Methods in Bifurcation Theory*, Lecture Notes in Mathematics 762, Springer-Verlag, Heidelberg, 1979.

[57] ———, *Bifurcation and symmetry breaking in applied mathematics*, Bull. Amer. Math. Soc., 3 (1980), pp. 779–819.

[58] ———, *Bifurcation from spherical symmetry*, AMS Summer Institute for Fluid Mechanics in Astrophysics and Geophysics, American Mathematical Society, Providence, RI, 1982.

[59] G. SCHWARZ, *Smooth functions invariant under the action of a compact Lie group*, Topology, 14 (1975), pp. 63–68.

[60] E. H. SPANIER, *Algebraic Topology*, McGraw-Hill, New York, 1966.

[61] R. TEMAM, *A nonlinear eigenvalue problem: The shape at equilibrium of a confined plasma*, Arch. Rational Mech. Anal., 60 (1975), pp. 51–73.

[62] ———, *Remarks on a free boundary value problem arising in plasma physics*, Comm. Partial Differential Equations, 2 (1977), pp. 563–585.

[63] A. WEINSTEIN, *Periodic orbits for convex Hamiltonian systems*, Ann. Math., 108 (1978), pp. 507–518.

[64] A. ZYGMUND, *Trigonometrical Series*, Dover, New York, 1955.